THE WATER CLOSET

A New History

THE WATER CLOSET

A New History

ROY PALMER

DAVID & CHARLES : NEWTON ABBOT

ISBN 0 7153 5763 8

Set in eleven on thirteen point Baskerville
and printed in Great Britain
by Latimer Trend & Company Ltd Plymouth
for David & Charles (Holdings) Limited
South Devon House Newton Abbot Devon

To Eileen with love

Contents

List of Illustrations

PLATES

In the text

(Figures 1, 2, 4, have been redrawn by David Evans from originals in an article by H. A. J. Lamb published in the *Architects' Journal*)

Introduction

THE development of the modern water closet and its attendant cistern and flushing apparatus is comparatively recent. Yet many previous civilisations have recognised the advantages of waterborne sanitation and archaeology has uncovered numerous attempts to solve one of man's ever-present, ever-pressing problems.

Sometime in the 1930s there was a small camp-site for cyclists and other wanderers in a remote part of North Wales. It consisted of little more than a field from which animals could be excluded and its only amenities were a ready supply of water and milk from a nearby farm and a curious wooden structure with the letters WC on its door. This simple building contained little more than a roll of paper and a seat with a hole in it. Beneath the seat was a strong flow of water from the stream over which the WC had been built. Here was a water closet at its simplest: an apparatus of continuous flush, entirely automatic, devoid of all smell and noisomeness. The pollution of the stream was another matter.

This structure merely shows how simple sanitation can be and presumably the first attempt at waterborne disposal occurred when some primitive man stuck his backside over a convenient stream. The next step might have been to have made a simple seat out of branches, and where water was plentiful and free-flowing this solution to the problem must have had its attractions.

Such crude examples have existed in many lands, at all stages of history, but because of their flimsy nature they have not survived for long. With urbanisation and the concentration of people into small areas came the need for more permanent structures, but the bedouin with thousands of square miles of sand at his disposal, or the North American Indian before the advent of the white man, had not the problem facing the citizen of London or New York, and indeed the primitive is still to be found, even in advanced countries. A terrifying example existed until comparatively recently in a Norwegian hill farm high up in the mountains, where the seat overhung a dizzy drop to the waters of the fjord 2,000 feet below.

So no inventor can be assigned to the privy. Early examples made of stone or other long-lasting material can still be seen by those curious enough to seek them out. The gradual evolution of todays' 'low-level flush suite' is here examined.

CHAPTER ONE

The Background and Chronology

EARLY historical evidence of how men dealt with a difficult problem over 2,000 years ago can be found in the Bible: for instance, Judges iii, 24, 'he must be relieving himself in the closet of his summer palace,' or 2 Kings x, 27, 'and they pulled down the sacred pillar of Baal and the temple itself and made a privy of it as it is today'. But such references are few, and it is more rewarding to examine the remains that have survived from earlier civilisations. The Ancient Egyptian, Indus Valley and Mycenean eras provide examples dating from before the birth of Christ; whilst Rome and Byzantium bequeath later, but still ancient, legacies.

One of the very earliest water closets or privies still surviving is to be found at Mohenjo-Daro in India. This dates back to the period 3250–2750 BC though there is some doubt as to the exact function of water in this case. Rai Bahadur Sahni, a former superintendent of the Archaeological Survey of India, wrote that in a house built towards the end of the period examples of two privies were to be found. They were connected to drains in the street by way of outlets in the walls. Sir John Marshall, one of the archaeologists engaged on the excavations, expressed the opinion that these privies were fitted with seats, 'as to the character of which there can be no question, and the drainage with which they are connected must have been designed for sewage of any kind, solid or fluid'. So the claim has been made that Mohenjo-Daro is the 'sanitary vanguard of mankind'.

Other early remains of privies date back to the Sumerians, whose civilisation began in the second half of the fourth millenium BC. The palace of their famous Sargon, King of Kings, revealed no less than six privies when it was excavated. They are of particular interest because of their essentially western design, having the high seats with which we are familiar today, instead of the customary Near East variety, which being level with the floor demands a squatting position by its users. The date of these would be about 2350–2130 BC.

One WC worthy of serious attention in any survey of their history was found in the Palace of Knossos in Crete, excavated in the first quarter of this century by Sir Arthur Evans, and still to be seen. Alongside the bathroom in the palace was a smaller room of which Sir Arthur wrote:

> On the face of a Gypsum slab . . . is a groove for a seat about 57cms from the floor. Outside the doorway of the latrine is a flag-stone sloped towards a semicircular hole, forming a sink, and from this opens a small duct leading to the main drain. The aperture from the main drain deviates from the centre of the seat, thus leaving room for some vessel used for flushing the basin.

Major rebuilding of the Cretan palaces following a disaster in the year 1650 BC enables this particular construction to be dated with some accuracy.

Ancient Egypt provides a magnificent example from the

eighteenth dynasty, *c* 1370–1350 BC, discovered in the house of a high official in Tel-el-Armana. The house contained a bathroom, and behind a screen was found a closet with a seat made of limestone, a very suitable material in a hot climate. This seat had been carved to an anatomical shape for the buttocks and is open on one side, thus anticipating the split seats of many modern WCS (see plate 1, p 17). There is no proof that this privy was flushed with water though the street outside had a stone drainage channel and a connection between the two may have existed.

In the forward march of civilisation it is never possible to overlook the Romans. Vespasian is remembered among other things for the construction of public urinals in the big cities. This might have seemed an act of public benefaction, but in fact the urine was collected in large vats for sale to the dye-makers, and was a source of income for the Emperor. Perhaps the finest monument to the efficiency of the Romans was that revealed by excavations at Timgad in North Africa. Here a room was discovered containing no less than twenty-five stone seats grouped round three sides of a room. Each seat was separated by a carving of a dolphin, then a popular decorative symbol.

As Rome declined and fell, Europe seems to have done little to improve sanitation, but with the rise of the Byzantine Empire a new interest seems to have sprung up, if not among the populace, at least among the hierarchy of priests, nobility and officers. In castle after castle which has been uncovered in modern times examples of WCS are to be found. In the ruins of a castle at Paphos in Cyprus are the remains of no less than four WCS. Two of these are back-to-back and were possibly separated by a wooden screen. The finest of the four consists of a cubicle built into the fabric of the walls, and contains a seat of golden limestone with a semicircular hole (see plate 2, p 17). At the base of the seat is a grooved channel, obviously used for flushing purposes. Outside the entrance are two curious depressions looking like soup plates, which may have held amphorae containing water. Other Byzantine sanitary structures are to be found at St Hilarion, near Kyrenia in Cyprus, where a castle is built on a mountain peak (this was the model for the castle portrayed by Walt Disney in his film 'Snow White and the Seven Dwarfs').

Beyond the Byzantine Empire the Christian countries saw an era of great decline and in the Middle Ages sanitation in England was almost non-existent. They could well be renamed the 'Midden Ages' for drainage and sewage accumulated in noisome pits or fly-ridden piles, bringing death and disease. The Black Death, which had reached its climax in England about the year 1350, served to highlight the needs of the people for efficient sanitation but little seems to have been done. In 1358 when the worst of the epidemic was spent there were only four public latrines in the whole of London. Even the largest, that at London Bridge, catered for the daily libations of 138 people only.

That middens and open cesspools were beginning to be of public concern is shown by a proclamation of 1388, one of the earliest on the subject:

> 12 Richard II. INFECTIONS
> For that so much Dung and Filth . . . be cast and put into Ditches, Rivers and other Waters, and also within many other Places, within, about and nigh unto divers Cities, Boroughs and Townes of the Realm, . . . that the air there is greatly corrupt and infect, and many Maladies and other intolerable Diseases do daily happen, as well as to the Inhabitants, Dwellers, Repairers and Travellers aforesaid. . . . Proclamation shall be made as well as in the Citie of London, as in other Cities, Boroughs and Townes, through the Realm of England, that all that do cast or throw any such annoyances, issues, dung, intrails or other ordure in Ditches, Rivers and Waters, he shall cause them to be

Plate 1 (*above*) An early Egyptian closet seat. This anatomically shaped seat was found in Tel-el-Armana and dates from about 1370 BC. The split nature of the seat anticipates modern versions

Plate 2 (*below*) An eighth-century closet seat of golden limestone from Paphos in Cyprus

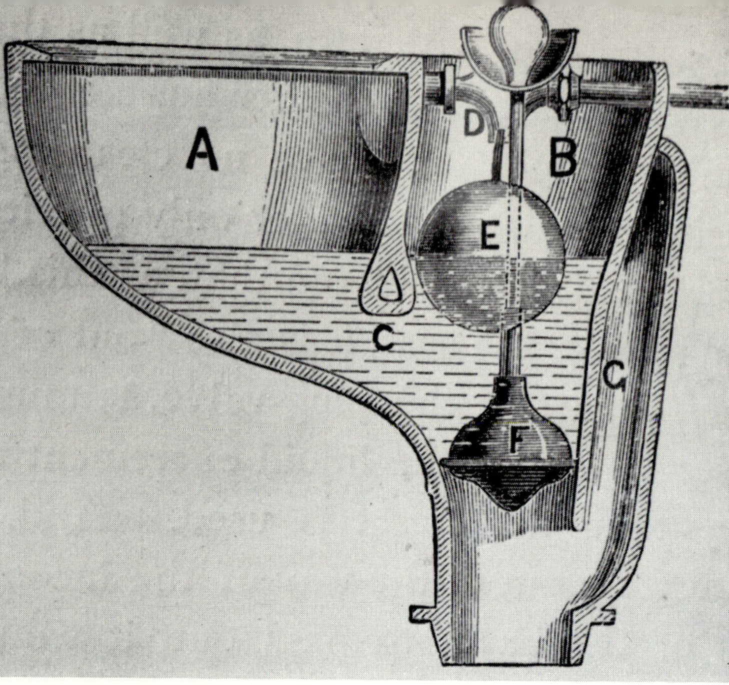

Plate 3 (*above*) The twin-basin closet was an early trapless basin which proved quite unsatisfactory for the job. It was impossible to keep clean

Plate 4 (*below*) The closet and trap in one piece manufactured by Jennings had many shortcomings

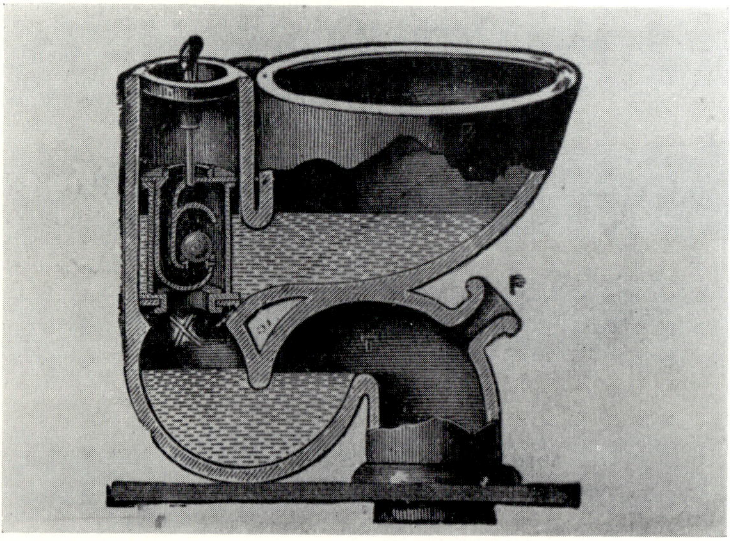

> removed . . . and carried away . . . upon pain to lose and forfeit to our Lord the King £20.

The story of sanitation seems to be one of spasmodic efforts soon forgotten. About 1449 Thomas Brightfield, a citizen of St Martins in London, conceived the idea of a stone privy which could be flushed with water from a lead cistern ingeniously constructed to catch rainwater. This clever arrangement does not appear to have spurred any contemporaries into copying it and the idea perished with its inventor.

There is little evidence, following Richard's proclamation, that the disposal of 'ordures, dung and annoyances' was much improved over the centuries. 35 Henry VIII c 15 went on at some length about the ancient town of Cambridge where the paving in both main roads and lanes were 'noyed with filth and mire' which lay in large mounds, a danger to the health of both inhabitants and travellers. Five years later, in 1549, the lack of a sewage system called for another Act, 3 and 4 Edward VI c 8:

> Wherein the Parliament held at Westminster, . . . in the XXII year of the most victorious reign of our late sovereign Lord King Henry VIII, among other things, one general act, concerning Commissioners of Sewers, . . . was enacted and made to continue for twenty years.

In 1596 there was a ray of hope when Sir John Harington, of Kelston near Bath, invented a valve closet with a flushing cistern. This important development which was the forerunner of the modern water closet will be dealt with in more detail in the next chapter.

Despite this, attempts to dispose of sewage material seem to have been irregular in the extreme. For the peasantry, and indeed for more elevated people, the injunction to retire 'a bow's shot' away was all the sanitary precaution they heeded. Though Acts of Parliament could often be ignored, the needs of certain communities ensured that some standards of sanitation were observed. In monasteries, nunneries, castles, convents and big houses the 'bow's shot' solution was inadequate. In 1662 'The Ancient Rites and Monuments of the Monastical Church of Durham' stated:

> There was also a large and decent place, adjoining to the West-side of the said Dorter, towards the Water for the Monks, and the Novices to resort to, called the Privies, two great Pillars of Stone bearing up the whole floor thereof. Every Seat, and Partition was of wainscot, closed on either side, so that they could not see one another when they were in that place.
>
> There were as many seats on either side as there were little Windows in the Wall to give light to the said Seats; which afterwards were walled up to make the House more close.

Customs had changed since the days of Timgad when defecation was regarded as an excuse for a social gathering in a public room, with a splashing fountain for company.

Another arrangement of some interest in the Middle Ages was that used by the monks of Tintern Abbey in Monmouthshire, who built a drainage system within flushing distance of the tidal part of the nearby River Wye. Convenient as this must have been for most of the year, there were times when the river rose above the desired height—with results that may well be imagined!

The nobility too endeavoured to introduce efficient sanitation into their medieval castles. Wherever a stream could be diverted it was channelled off to the so-called 'garde-robe'; numerous examples of these exist in castles scattered throughout the British Isles. Langley Castle has four closets on each floor of the tower emptying by separate flues into the stream below. At Chepstow Castle a privy overhung the river 200 feet beneath. Where a stream was not available the moat was often used instead, as at Pembridge in Monmouthshire where two closets overhung the water.

In country houses the garden was often the recommended place for anyone to go and 'pluck a rose'. Jonathan Swift, writing in a pamphlet, 'Directions to Servants' (1745), was bitter in his condemnation of women who would not seek to relieve themselves in the garden:

> I am very much offended with those Ladies, who are so proud and lazy, they will not be at the Pains of stepping into the Garden to pluck a Rose, but keep an odious Implement, sometimes in the Bed-chamber itself, or at least in a dark Closet adjoining, which they make use of to ease their worst Necessities, and, you are the usual Carriers away of the Pan, which maketh not only the Chamber but even their Cloaths offensive, to all who come near. Now, to cure them of this odious Practice, let me advise you, on whom this Office lieth, to convey away this Utensil, that you will do it openly down the great Stairs, and in the Presence of the Footmen; and if any Body knocketh, to open.

Notwithstanding Swift's conviction that the garden was a suitable place the fact was that many houses, especially in large towns, simply did not have gardens attached. Thus arose the custom of rubbish of all kinds being ejected from upstairs windows to the peril of anyone walking beneath. Though banned as early as 1395 in Paris, the practice was still observed in Edinburgh in 1750. The classical cry of 'Gardez l'eau' was not always forthcoming and S. S. Hellyer writes about it in his book *Plumbing* (1893):

> . . . when people went into the streets at night, it was necessary, in order to avoid disagreeable accidents from the windows, that they should take with them a guide, who, as he went along, called out with a loud voice 'Haud your hand'. This must have been a good time for hatters and tailors. At that time when the luxury of water closets was unknown, it was the custom of men to perambulate the streets of Edinburgh carrying conveniences (pails) suspended from a yoke on their shoulders, and enveloped by cloaks sufficiently large to cover both their appliance and customers, crying 'Wha wants me for a bawbee?' It has since been used against the Edinburgh people as a joke or satire upon

an ancient custom. By way of set-off, it may be observed, that in 1846 almost every house in Edinburgh had a water closet.

Yet in 1750 Edinburgh, like many other towns, had not much longer to wait. In 1775 Alexander Cummings, of London, took out the first patent for a water closet. This was followed by another patent by Samuel Prosser in 1777 and by Bramah in 1778. These will be considered in a later chapter.

By the turn of the century water closets were becoming more familiar objects in the English home but there were many places where the old practices lingered on. In 1832 Exeter was described as having open gutters as the only means of disposal of sewage and household slops. Nor was Nottingham any better. There the inhabitants were in the habit of emptying 'chamber vessels into the grates over the sewers'.

From about 1825 there are increasing references in print to WCS. *Kelly's Price-Book* of that year advertised a valve closet, complete with all appurtenances for £6 6s. A less expensive model with a cheaper pan was available at £3 10s.

If any one event in the nineteenth century can be singled out as popularising the WC it is, without doubt, the publicity that followed the publication of *The General Report on the Sanitary Conditions of the Labouring Classes in Great Britain* in 1842. This report was written by Edwin Chadwick, to whom scant praise has been given by many historians of the last century in England. R. A. Lewis, in his book about Chadwick, says: 'Few men have done so much for their fellow-countrymen . . . and received in return so little thanks.' Perhaps this was Chadwick's own fault; it seemed that he was an impossible man to like, with his arrogance, ruthlessness and uncomprising attitudes.

The report of 1842 which bears his name is too extensive a document to be dealt with fully here and concerns WCs only incidentally. It brought into sharp focus the large amount of preventable illness, typhus, dysentery, cholera and other waterborne infections suffered by large sections of the British population to whom sanitation, adequate sewage facilities and clean water supplies were unavailable or scanty. Among other things Chadwick instructed John Roe, Surveyor to the Holborn Sewers Commission, to undertake experimental work on the problem of providing glazed pipes for the disposal of sewage in place of the old brick drains. This in itself was a revolution, as we shall see later.

Chadwick believed that for as little as £4 a house could be fitted with a WC, good drains and a sink, but he made many enemies with his ruthless pursuit of the 'sanitary idea'. His report dropped like a bombshell—at last people were made to realise the extent of the sanitation problem and to demand action. The need for plentiful water supply, efficient drains and water closets could no longer be overlooked. At last the WC had ceased to be the prerogative of the upper classes and had become a necessity for all.

As a result of the Chadwick Report and the public outcry a Public Health Act was passed in 1848. Among its many clauses was one which made it obligatory for all new houses to be provided with a WC, privy or ash pit. Nor was it only in London that the new-fangled WCs were being installed. Thus Iredale, writing in *The History of Congleton*, says that after about 1848 the demand for good plumbing and sanitation increased: 'As early as that Pedley and Massey . . . were advertising their new self-acting closets.'

In 1854 the Sanitary Amendment Act empowered the Medical Officer of Liverpool to compel owners of houses to install water closets in homes where health hazards existed.

At long last, by an Act of 1875, sanitary facilities were made compulsory in all new housing (38 and 39 Victoria c 55):

> It shall not be lawful newly to erect any house, or to rebuild any house pulled down to or below the ground floor, without a sufficient water closet, earth closet or privy and an ashpit furnished with proper doors and coverings.

In 1871 Chrimes took out three patents for new and improved ball valves. There was also a bewildering proliferation of water closets, cisterns and fittings during this inventive period. Sanitary conditions were continually improving, though in many places a great deal of work was involved. Thus in Manchester in 1881 no less than 44,000 primitive 'cinder-sifter' closets remained. In nearby Burnley with a population of 100,000 living in 20,000 houses there were, in 1888, 6,586 wcs and 9,009 pail privies. By 1895, however, the wcs had increased to 16,842 and the pail privies had fallen to 3,816. The Sanitary Inspector was able to say with some confidence, 'Within a few years it is expected that every town will have waterborne sanitation.' In 1896 Donald Cameron claimed the discovery of the septic tank, which in the words of the *Exeter Flying Post* would spread 'the name of Exeter to the ends of the earth'.

Across the Atlantic the developing United States were equally alive to the benefits of sanitation. In 1847 Thomas Maddocks had landed in New York from England and by 1873 was selling his American-made sanitary equipment. A scrutiny of patents in the USA reveals that the first one was for a plunger closet taken out by William Campbell and James T. Henry in 1857. Such closets resembled in many respects the twin-basin wcs (see plate 3, p 18), deplored by S. S. Hellyer in his writings for their insanitary mechanism, and the trapless closet of Jennings (see plates 4 and 5, pp 18 and 35). However, the latter patented a plunger closet as late as 1876 and it was still further developed by W. S. Carr and Demarest, both American patents. Initially the Americans had looked to Europe, and particularly England, for the supply of their sanitary needs. In *The Sanitary Engineer* of 9 August 1883, a long article was published eulogising the sanitary arrangements in a London restaurant. Lavatory basins and urinals are described, and full details are given of the water closets, in particular the bends and traps on these which were described as of 'the American drawn-lead pattern'.

So the story continued, each advance bringing ever higher standards and advantages. Then in 1907 the water closet was defined by Act of Parliament:

> The expression 'closet accommodation' includes a receptacle for human excreta, together with a structure comprising such receptacle and the fittings and apparatus connected therewith. . . . The expression 'water closet' means closet accommodation used or adapted or intended to be used in connection with the water carriage system, and comprising provision for the flushing of the receptacle by means of a fresh water supply, and having proper communication with a sewer.

Thus the water closet became an accepted part of the apparatus of the home, and a necessary part of hygienic living. It is remarkable that it took so long for such a basic piece of equipment to evolve, and equally remarkable that it was adopted on such a large scale within a century, so that now we take it for granted. Indeed we do more than this: we try to deny its existence by seeking euphemisms for its name.

CHAPTER TWO

Sir John Harington

As we have seen many attempts were made as far back as the days of Mohenjo-Daro and Knossos to provide waterborne sanitation for at least some favoured members of the community. It was not, however, until the sixteenth century that a cheap and practical apparatus was invented, one moreover that initiated many of the devices that are in common use today.

The inventor of this remarkable machine, Sir John Harington, was a kinsman of Queen Elizabeth I. He not only built the first known valve closet, but he wrote a book of great wit and learning about it and gave full details to anyone wishing to emulate him. Few did, and the water music that Sir John started was not to be heard again for nearly two hundred years.

The Harington family had originated in the small Cumberland village of Hudleston, where their fortunes alternately ebbed and flowed with the Wars of the Roses. An early forbear, Sir James, captured Henry VI and was rewarded with lands when Edward IV became king. With the defeat of Richard III in 1485 at Bosworth, and the subsequent accession of Henry VII, the tables were turned and the family suffered a decline in wealth and prestige. Between 1485 and 1561 when John Harington was born, the family moved to Stepney. By now the family had retrieved its former status and Queen Elizabeth, who had then been on the throne of England for three years, graciously consented to be godmother to the new infant. So it was that the young Harington was familiar in court circles and was surrounded by a host of influential friends. His fortunes tended to rise or fall as he incurred the queen's pleasure or her anger. There were times when her frown cast a cold shadow over him, one occasion arising from his translation in 1584 of a scurrilous story of two cuckolds from Ariosto's *Orlando Furioso*. John was foolish enough to circulate this *risqué* story among the maids-of-honour at court with the inevitable result that it fell at last into the queen's hands. Like Victoria many years later, she was not amused, and he was banished from court. By 1591 he was back in favour again, his sins purged by further wit, this time of a less racy flavour.

Meanwhile he had married Mary Rogers, who had brought as part of her dowry the estate of Kelston, a small village near Bath. During the years of exile from court the old house at Kelston had been pulled down and a new one built to the design of the famous Italian architect Barozzi of Vignola. It was in this new house that the now famous WC was constructed, a device that was later to know the royal presence. Now this house too has gone and with it the closet, but the family continued to live in Kelston for a considerable time and there are many memorials to the Haringtons in the nearby village church (see plate 6, p 35).

In 1592 Sir John became High Sheriff of the county of Somerset and in that official capacity he entertained the queen when she made a visit later that year. What demonstrations and conversation took place it is impossible to guess. Suffice it that the queen was sufficiently impressed by Harington's expertise to have a similar WC installed for her own personal use in Richmond Palace.

These are the bare facts. Harington had other claims to fame but they are not of importance here. What is really of note is the 'fantastical treatise' he wrote concerning his device. Whether this book was an altruistic attempt to pass on the benefits of his labours or merely self-advertisement may be judged by his remark: 'I was the willinger to wryte . . . because I thought this would give me some occasion to have me thought of and talked of.'

Whatever the motives, the book was written, full of humour and practical advice. It was published by Richard Field under the title of *A New Discourse of a Stale Subject; Called the Metamorphosis of Ajax, Written by Misacmos to his Friend and Cousin Philostilpnos.*

Early on in the book is one of Harington's attempts at poetry which reveals him as a man of letters, though he has been unjustly described as 'like Sterne at his worst'.

A goodly Father sitting on a draught,
To doe as need, and Nature hath us taught,
Mumbled, as was his manner, certaine prayers:
And unto him, the Divell straight repaires,
And boldly to revile him he begins,
Alleaging, that such prayers are deadly sinnes;
And that it proved he was devoid of grace,
To speak to God in so unfit a place.
The reverend man, though at the first dismayed,
Yet strong in faith, thus to the Divell said:
Thou damned Spirit, wicked, false and lying,
Despairing thine owne good, and ever envying:

Each take his due, and me thou canst not hurt,
To God my prayer I meant, to thee the dirt.
Pure prayer ascends to him that high doth sit.
Downe falls the filth, for fiends of hell more fit.

For the student of the WC the importance of the book is that it gives the practical man advice for the construction of a machine himself. The printed word is accompanied by sketches which exhibit a curious perspective but embellish a reasonably self-explanatory text. In it a fictitious character who signs himself 'T.C.'—a traveller—directs a letter to 'M. E. S. Esquire':

Sir,

My master having expressly commanded me to finish a strange discourse that he had written to you, called the Metamorpho-sis of Ajax, by setting certain pictures thereto. . . . Wherefor now to instruct you and all gentlemen of worship, how to reform all unsavoury places of your houses, whether they be caused by privies or sinks, or such like (for the annoyance coming all of like causes, the remedies need not be much unlike) this shall do.

AN ANATOMY

In the privy that annoys you, first cause a cistern, containing a barrel, or upward, to be placed either in the room or above it, from whence the water may, by a small pipe of lead of an inch be conveyed under the seat in the hinder part thereof (but quite out of sight); to which pipe you must have a small cock or washer, to yield water with some pretty strength when you would let it in.

Next make a vessel of an oval form, as broad at the bottom as at the top; two feet deep, one foot broad, sixteen inches long; place this very close to your seat, like the pot of a close-stool, let the oval incline to the right hand. This vessel may be of brick, stone or lead; but whatsoever it is, it should have a current of three inches to the back part of it (where a sluice of brass must stand); the bottom and sides all smooth, and dressed with pitch, rosin and wax: which will keep it from tainting with the urine.

In the lowest part of the vessel which will be on the right hand, you must fasten the sluice or washer of brass, with solder or cement; the concavity or hollow thereof, must be two inches and a half.

To the washers stopples must be a stem of iron as big as a curtain rod; strong and even, and perpendicular, with a strong screw at the top of it; to which you must have a hollow key with a worm fit to that screw.

This screw must, when the sluice is down appear through the plank not above a straw's breadth on the right hand; and being duly placed, it will stand about three or four inches wide of the midst of the back of your seat.

That the children and busy folk disorder it not, or open the sluice with putting in their hands without a key, you should have a little button or scallop shell, to bind it down with a vice pin, so as without the key it will not be opened.

If water be plenty, the oftener it is used and opened, the sweeter; but if it be scant, once a day is enough, for a need, though twenty persons should use it. . . . And this being well done, and orderly kept, your worst privy may be as sweet as your best chamber.

But to conclude all this in a few words it is but a standing close-stool easily emptied. And by the like reason (other forms and proportions observed) all other places of your house may be kept sweet.

Your worships to command,
T.C.,
Traveller.

When later a WC was fitted for the Queen's pleasure in Richmond Palace, Harington was once more moved to write. A copy of Ajax was chained to the wall and the following epigram directed to the ladies of the Privy Chamber:

Faire Dames, if any look in scorn, and spites
Me, that Misacmos Muse in mirth did write,
To satisfie the sinne, lo, here in chains,
For aye to hang, my Master me ordaines.
Yet deem the deed to him no derogation,
But deign to this device new commendation,
Sith here you see, feele, smell that his conveyance,
Hath freed this noysome place from all annoyance.
Now judge you, that the work mock, envie, taunt,
Whose service in this place may make most vaunt:
If us, or you, to praise it, were most meet,
You that made sowre, or us, that make it sweet?

Alas the book did not remain 'for aye to hang'. The wcs both at Kelston and Richmond disappeared, sunk like Atlantis beneath the waters. The 'Midden Ages' wallowed on until Cummings re-invented the device in 1775. One wonders if he really invented it, or whether by chance he came across a copy of Ajax and adapted its contents for his own use?

CHAPTER THREE

Types of Water Closets

It was not until the idea of waterborne sanitation became accepted by the public that improvements and inventions really began to follow earlier imperfections. The Chadwick Report of 1842 and the Health Act of 1875 were, as we have seen, responsible for giving impetus to an already accelerating movement. Soon many changes began to take place on both sides of the Atlantic.

From its beginning until comparatively recently the WC has known at least six basic types, many of them being manufactured at the same time. An illustration of this is the statement by the famous Victorian plumber, S. S. Hellyer: 'There is no WC equal to a good valve WC', when Twyfords had been making a wash-out closet for the previous seventeen years. Early records are difficult to trace and when found are often contradictory, thus it is difficult to designate the originators of much work. Some facts capable of verification emerge towards the end of the nineteenth century and will be dealt with in due course. Early WCs had simple pans of metal or earthenware and it was not until the increasing skills of the potter enabled more complex bowls to be made that the modern type of WC, such as we know today, began to appear. It is convenient to recognise the following types:

1 The Pan Closet

This highly objectionable closet consisted of an upper earthenware basin with a shallow copper pan containing three to four inches of water as a seal at its base. This pan could be tipped away to discharge its contents into a lower large cast-iron receptacle connected to the drainage system (see plate 7, p 36) S. S. Hellyer, writing in 1890, says that although it has

> been improved many times by various manufacturers, one wonders what it could have been like before William Law 'improved' it in 1796, for it remains to this day, . . . about the most insanitary closet in use. . . . The only 'bliss' that the public can have about so foul a thing is ignorance of its nature.

Hellyer was not alone in his criticism of the pan closet. P. J. Davies, a practising plumber of the nineteenth century had this to say of it:

> I consider this closet a very unsanitary piece of mechanism, and totally unfit for its intended purpose, inasmuch as in a short time the internal parts become besmeared, and consequently become offensive.

Despite such outspoken condemnation by two such figures as Hellyer and Davies, many were still in use in the twentieth century in London and other parts of England. Thousands too had been exported to Russia.

In the earliest days of waterborne sanitation the Americans had been content to import the insanitary pan closet from

Great Britain, and it is interesting that no US patent for this device exists. Certain pan closets with claims to be 'patented', notably those of Carr, Bartholomew and Harrison, were really improved versions of closets of British origin with patented features relating to the shape of the pan, more efficient flushing devices, lever mechanisms and so on.

Attempts had been made by Banner to remove some of the objectionable features of the pan closet and the so-called 'Banner System' was the result of his thought on the subject. His thesis was that no pan closet should have an individual trap below it, contrary to the beliefs of most authorities, and that the system should have instead an interceptor trap at the base of the soil pipe where it entered the drain to the sewer. Above this trap was a fresh-air inlet which kept the soil pipe above the trap full of fresh air. The soil pipe also connected with a cowl on the roof of the house. Davies was not impressed:

> But whatever Mr Banner's views may be about fixing a closet without a trap immediately below it, I certainly, for one, can't cotton on to his ideas; . . . and with all due deference to Mr Banner I say that no closet of any kind yet in the market can be considered safe without it.

2 The Valve Closet

This was Harington's invention as has already been noted. It operated on a system by which a pan with an opening at the bottom was sealed by a leather-faced valve. When flushing took place a complicated arrangement of handle, lever and counter-weights admitted water to the bowl whilst temporarily removing the valve. Owing to this cumbersome mechanism, which occupied a lot of room, the machine was usually camouflaged with a wooden surround of varying quality and ornamentation.

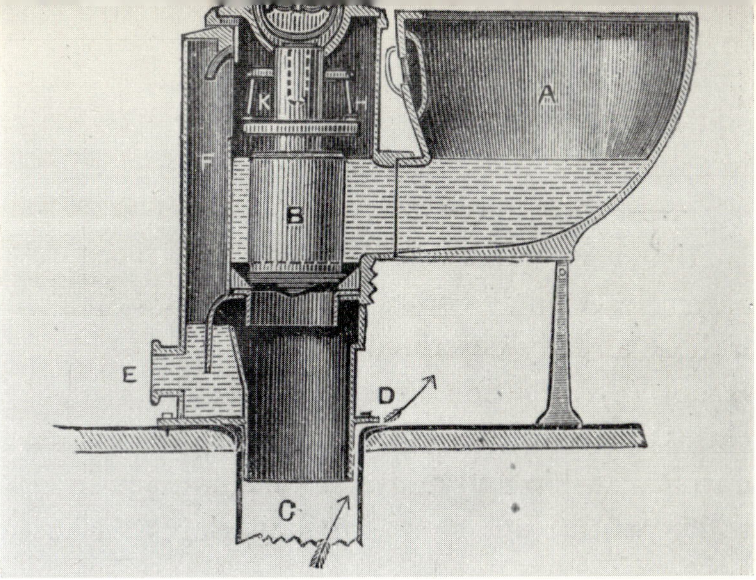

Plate 5 (*above*) Jennings' trapless closet was another unsuccessful attempt to produce an acceptable piece of apparatus. Like the twin-basin closet it was difficult to keep clean

Plate 6 (*below*) One of the earliest builders of a water closet was Sir John Harington. The memorial to one of his descendants shows clearly that the family originated from Hudleston in Cumberland

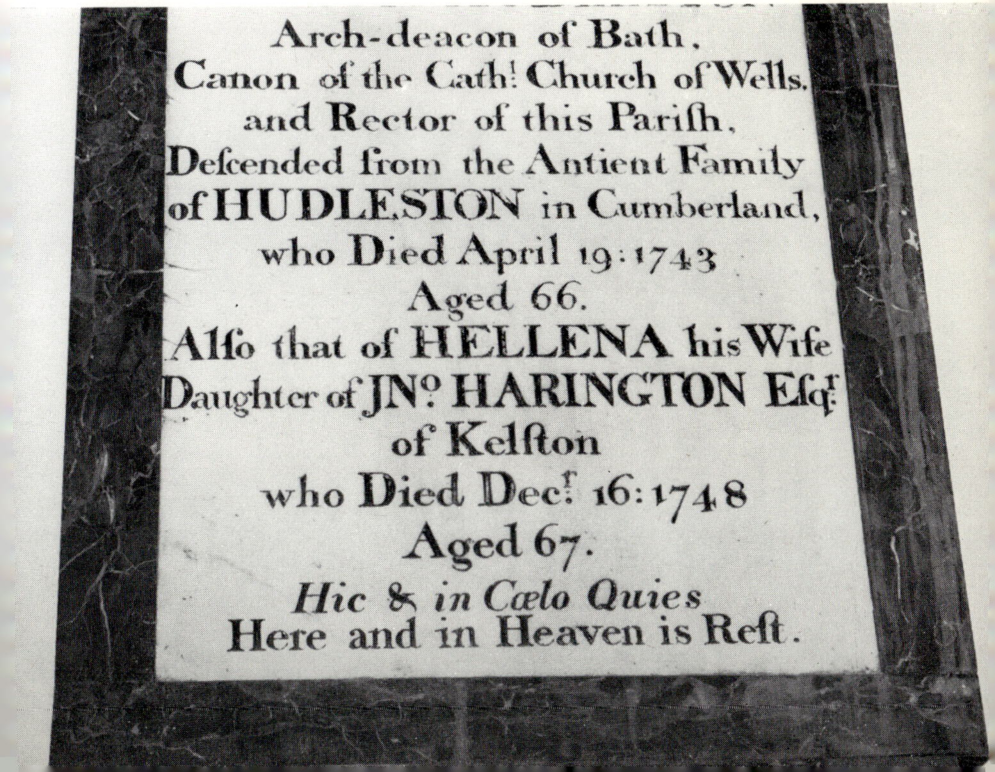

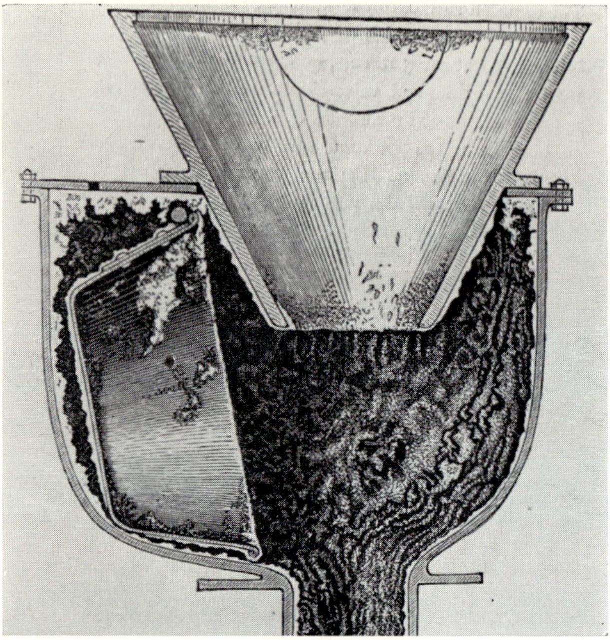

Plate 7 (*above*) 'About the most insanitary closet in use,' was one expert's opinion of the pan closet. The photograph reveals why

Plate 8 (*below*) This exposed Doulton valve closet shows the need of such early closets for a wooden surround

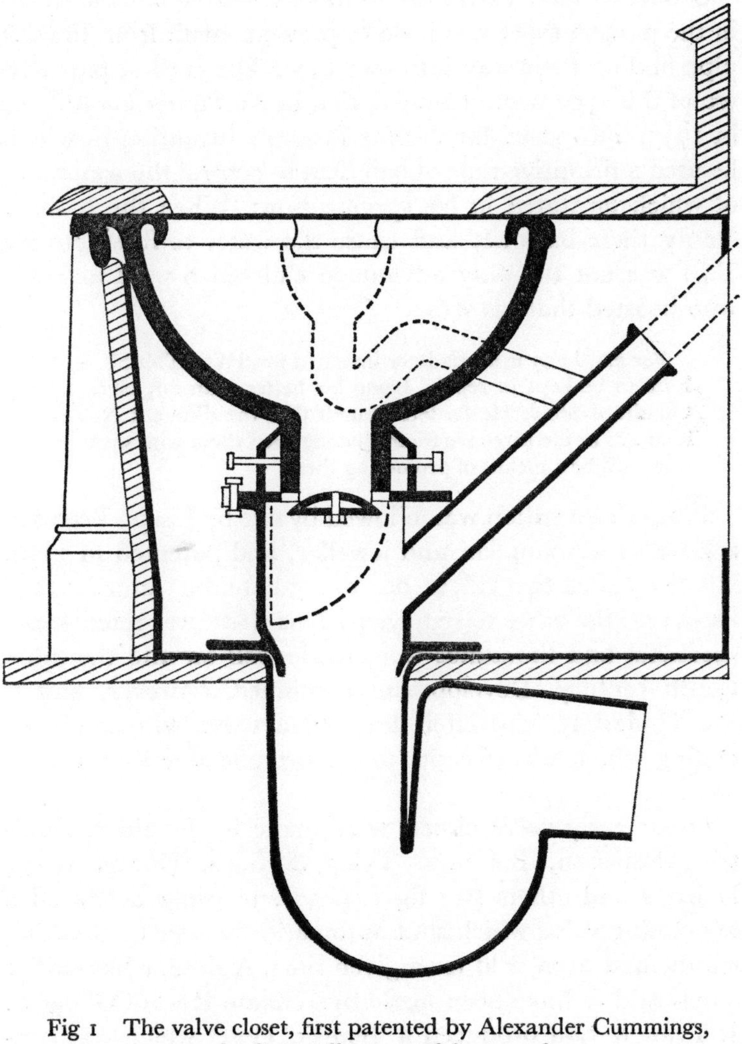

Fig 1 The valve closet, first patented by Alexander Cummings, was the first really successful water closet

Sooner or later valves began to leak, letting out the water in the pan, so there was little to prevent smells from the soil pipe finding their way into the room. The earliest patented WC of this type was, of course, that of Alexander Cummings in 1775. Two years later came Prosser's invention in which he used a primitive type of ball float to control the water and of which he stated in his specification: 'When the basin is empty these balls act and cause the water to flow into it.' This was not the only advantage claimed by the inventor who boasted that his WCs

> Far excel any ever made or invented for SWEETNESS, and Ease to be kept in repair, being far better than any Slide or Quadrant Stool. He flatters himself that the different Noblemen . . . in the three kingdoms having used them with satisfaction, will be a means of promoting them.

Prosser's invention was followed by one by Joseph Bramah, a London watchmaker and jeweller, and patented in 1778. His closet used two valves, one to regulate the water into the basin and the other to remove it. This closet was remarkably successful and thousands were made well on into the nineteenth century. Development continued, however, and it was Underhay who later designed a valve with a rubber seating which was cheaper and more effective than earlier ones.

Other early valve closets were made by Jennings, Doultons, Smeaton, Boldings, Tylor & Sons, Humphersons, Hellyers and others (see fig 1, p 37). Jennings produced a self-closing valve which shut automatically after the flushing handle had been held for a given time. A similar but earlier one is said to have been made by William Ross of Glasgow. J. Tylor & Son produced a 'Bourne' closet which was considered 'one of the best in the market, every portion being strong and serviceable'.

Perhaps the most famous of all valve closets was the 'Optimus' made by Hellyers of Red Lion Square, London WC [sic]. In 1892 S. S. Hellyer was claiming many improvements, protected by letters patent, among which he listed:

(i) A flushing rim to the basin. 'There was no valve closet with a flushing rim to be had in the United Kingdom or in America . . . before I introduced it.'

(ii) Elastic valve for ready changing.

(iii) Table-top and basin in one piece.

3 The Hopper Closet

It used to be customary in the early days of wcs to select pan closets for the gentry and hopper closets for the servants; thus implying that the excretory functions of the upper and lower classes differed.

The hopper was a grossly ineffective wc consisting of a 'long' or 'short' basin, almost like a dunce's cap upside down (see plate 9, p 53). In the nineteenth century such a basin could be bought for 2s 6d, while the wash down (qv) would be five times as much. This crucial fact does much to explain the popularity of hoppers, which were turned out in thousands despite their shortcomings.

Flushing was by means of a meagre spiral squirt of water, which being little more than enough to wet the basin was quite inadequate to clean it.

4 The Wash-Out or Flush-Out Closet

Despite the name 'wash out', this was a highly effective and popular piece of apparatus and the forerunner of the modern wc. The idea of a valve to close the bottom of the pan was abandoned and instead a small amount of water

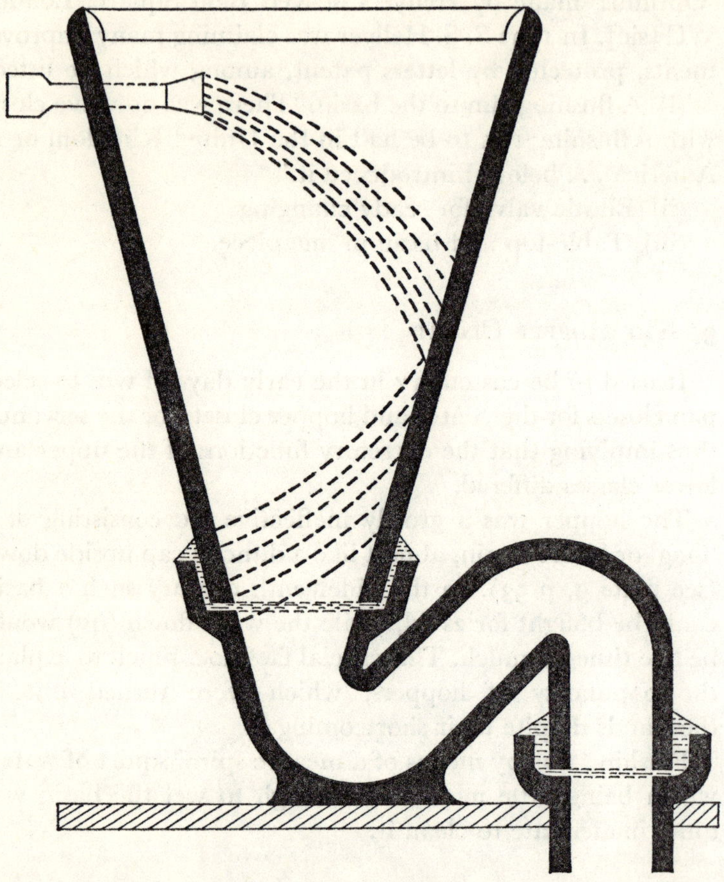

Fig 2 The 'hopper' was condemned on hygienic grounds by many leading sanitarians. The feeble nature of the flush can be understood from this drawing

which acted as a seal was left after each flush. Such wcs were soon in great demand and Twyfords, Doultons and Humphersons were early pioneers in this field. By 1889 Twyfords were claiming that 100,000 of the 'National' were in use. Dr Robert Park, in a letter entitled 'Sanitation at the Sea Coast' wrote: 'Of wash-out closet basins there are now several in vogue, but the best I have seen are those called Twyford's National and Alliance and a cheaper sort on the same principal called the Crown.' Prices ranged for the National from £1 10s in white to £2 2s in printed flown colours. Each closet was tested at the works with a 2 gallon cistern and was guaranteed to work satisfactorily.

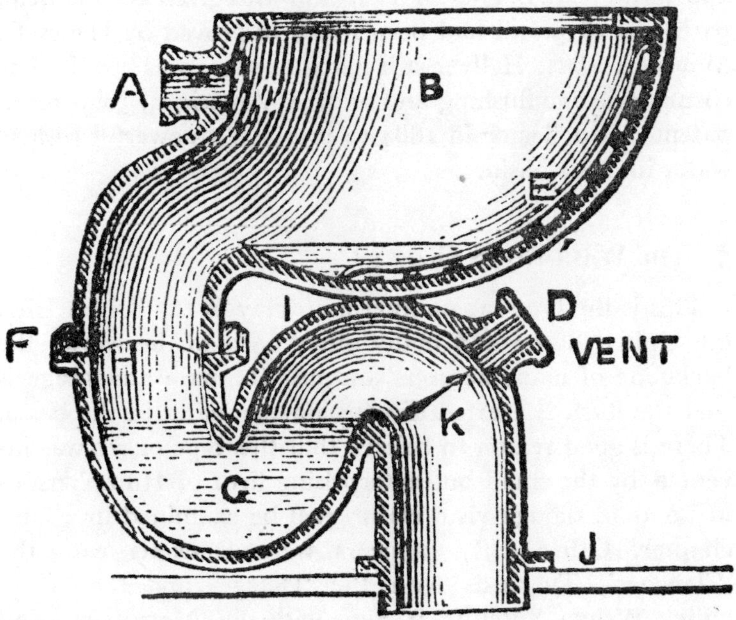

Fig 3 P. J. Davies the famous Victorian plumber, believed this to be the first wash-out closet made. Unfortunately no date was given

It is difficult to discover precisely who first conceived of the idea of the wash-out closet and further research is necessary. P. J. Davies in volume I of his *Standard Practical Plumbing* has this to say of one of his illustrations, a most obvious wash-out type: 'This represents one of, if not absolutely the first made, and is a very good one for the reason that the basin is constructed so that there is always a layer of water for the soil to drop into.' Unfortunately neither the date nor manufacturer of this closet is given and the statement must be regarded with some reservations (see fig 3, p 41).

The wash-out models proved to be as popular in the USA as they were in Europe and several modifications were to appear there. S. S. Hellyer's 'Artisan' was given an American patent in 1879 and had been much improved by Henry C. Meyer by 1881. Hellyer still felt that more was required to ensure effective flushing and another American, John Reid, patented a deflector in 1884 to produce a powerful rush of water into the basin.

5 The Wash-Down Closet

This is the type in common use today in the British Isles, the United States and many other parts of the world, the backbone of manufacturers' output. Its main advantage is that the flush is exerted directly on the water in the basin. There is good reason to believe that this type of WC was invented by the small but enterprising firm of Humphersons in London, though rival claims will be examined in a later chapter. Other early pioneers were Ducketts with the 'Clencher', Twyfords with the 'Deluge' and the 'Twycliffe', Wright Sutcliffe & Sons with the 'Metropole', and others.

6 The Siphonic Closet

The water and contents in the basin of this wc are carried away by means of a siphonic action. The earliest example, invented by John Randall Mann in 1870, had three pipes which delivered water into the basin. One pipe fed the flushing rim round the edge of the basin; a second discharged about half a gallon rapidly on to the water already in the basin, and started the siphonic action; the third produced a slower flow for an after flush (fig 4, p 51). Twentieth-century models (plates 22 and 25, pp 108 and 110) show how much the siphonic closet has been improved.

The first development of a siphonic closet in the USA was under the direction of William Smith in San Francisco. His first patent, issued in 1876, dealt with the emptying of the basin with a jet initiating siphonic action. Smith's principle was further developed by the famous American sanitarian Colonel George E. Waring with the later production of larger and more complicated pieces of sanitary ware.

In 1888 another American, Thomas Kennedy, brought out a siphonic closet which required two delivery pipes—one to flush the rim, the other to start the siphon. This wc was improved by William Howell in 1890 when he eliminated the lower trap without detriment to its action.

In the same year Robert Frame and Charles Neff of Newport, Rhode Island, produced what they hoped would be the solution to the siphonic principle. It was unfortunate that their closet sometimes failed to develop the necessary action with the result that the contents of the basin overflowed. Despite this the Frame and Neff closet was the prototype from which sprang today's American siphonic wash-down closets. Further improvement came when Fred Adee redesigned the shape of the bowl making it more efficient. In

1902 Frank A. Wells carried the process a stage further and the principle of siphonic water closets had been solved in America.

A great disadvantage of early siphonic closets arose from the common Victorian practice of throwing slop water down them. This was often enough to start the siphonic action without the use of the overhead cistern, with the danger that the trap in the closet would be emptied. The concern of plumbers in this matter was expressed by P. J. Davies writing of his own firm's 'Kensington' which 'depended upon the main-trap, and cannot be emptied out or the seal broken during the time the trap is at rest'.

Though the above cover the main types of WCs which found favour during the nineteenth century this by no means exhausts the list. Jennings at one time marketed a spray and jet pedestal closet which was meant to serve the double function of closet and bidet. A spray supplied with hot or cold water could be slid into place over the WC basin enabling the lucky owner to complete his toilet with a cleansing jet.

Another interesting development was the ship closet made to function below the waterline. This worked by pneumatic action foreshadowing the vacuum closet of the twentieth century. Early yachts to be fitted with this closet were the *Valhalla*, *Thistle* and the *Charmion* but there were many others. Similar WCs were made for railway trains.

There were even portable WCs with water-storage facilities which were claimed to be entirely inoffensive. Emptying them must have proved an unpleasant task for any servant.

A further late-nineteenth-century invention of Jennings was a three-in-one apparatus which combined a commode, a lavatory (or wash-hand basin) and a bidet—a truly fearsome piece of apparatus that was probably never very popular. It seems that the only wc left to be invented was that for the aeroplane, and that had to wait for the invention of the aeroplane itself!

CHAPTER FOUR

Some Sanitary Pioneers

THE water closet which started its official life in 1775 with Alexander Cumming's patent has seen many developers and improvers. Among them has been the firm of Hellyers (later Dent & Hellyer) in Victorian London. Claiming to have served the public through seven reigns, it was really put on the map by S. S. Hellyer whose remarkable inventive genius was brought to public notice by his lectures, books and pamphlets on sanitation. His most famous publication *The Plumber and Sanitary Houses*, subtitled *Water, Air, Light,* is now a collector's item. First published in 1877 it is one of the best nineteenth-century books on its subject and had reached its fifth edition in 1893. The fourth edition was translated into French under the patronage of La Chambre Syndicate des Entrepreneurs de Plomberie de la Ville de Paris. This serves to remind us that in the nineteenth century Britain probably led the world in the matter of the WC. The dedication of the first edition of Hellyer's book reads:

To the
DESIGNERS AND BUILDERS OF HOUSES,
HEAD-WORKERS AND HAND-WORKERS,
THIS LITTLE WORK
IS
WITH ALL DUE RESPECT,
Inscribed
BY THEIR OBEDIENT SERVANT,
S. STEVENS HELLYER.

Remarkable as Hellyer's accomplishments were in the plumbing world, his observations are still worth reading on their own account: 'There are a "thousand gates to death"! Fewer are wider, or open more readily, than those in our own homes, when unlocked by noxious gases or bad air from drains.'

He discusses at great length in his books the size of cisterns and the restricted use of water which spurred him to passionate anger. Some of his important observations can be found in the chapter on 'Seats and Cisterns'. He was also lyrical on the subject of the 'hand-worker':

> It is important that the head-worker who plans a work should understand what he plans; but it is equally important that the man who does the work—the hand-worker—should understand it too.
>
> Educate the hand worker should be the cry of all who wish for true progress in sanitary matters.

In another context S. S. Hellyer describes an event which has often been quoted:

> Harper's Magazine reviewing my 'Lecture on the Science and Art of Sanitary Plumbing' states that H.R.H. the Prince of Wales said, when recovering from a fever which was nearly fatal 'if I were not a prince, I would be a plumber'. This illness of the prince, (perhaps from bad plumbing) has done more for the advancement of the principles of sanitary plumbing than could have been accepted by being a member of the craft.

There follows a long diatribe on the position of closets, of bad drainage and the contamination of drinking water and other water. He then continues with another strong condemnation, this time of dustbins.

Notwithstanding all these sanitary publications S. S. Hellyer was not unaware of the value of regular trade catalogues to advertise the wares of his firm. His especial pride and joy was the 'Optimus' in all its variations. In essence the 'Optimus' was little different from the older 'Bramah' and was a valve closet destined to be made obsolete by the newer and more efficient wash-out and wash-down kinds. In Hellyer's hands the valve closet reached perhaps the zenith of its development and its popularity continued long after its demise might have been expected. As late as 1907 Dent & Hellyers were advertising that valve closets of the 'Optimus' type had been fitted in a number of famous places, including the Houses of Parliament, the Royal Courts of Justice, the Foreign Office, the Admiralty, the Jockey Club, the National Liberal Club and Hurlingham. They had also been exported to India, Australia, Natal, France, Italy, Spain, Germany, Austria, Hungary, Sweden, Denmark, Belgium, Holland, Switzerland, Turkey, Greece, South America, China, Japan and the USA. The king lowered the regal backside on to an 'Optimus' both at Windsor and Balmoral, setting precedence for HRH the Princess of Battenburg and the lesser rear ends of the dukes of Norfolk, Richmond and Gordon, Marlborough and Portland. It was even reported that the Duke of Bedford had taken one to sea in his yacht.

There were lots of models to choose from. In 1913 Hellyers produced an all embracing catalogue with something for everybody. One was advertised as having a slop top and was most suitable for ladies' use as 'the unpleasant draught experienced when using enclosed closets of the ordinary kind is not felt'.

Another had a projecting slop top with a curved front. More room was thus afforded the feet and the user could sit well back and with more comfort 'than is possible with a closet having a straightriser'.

There were spartan models as well. The 'Jar' was a patent wash-down closet made of fireclay and designed for use in positions where whiteware models would be liable to breakage from rough usage or frost (see plate 10, p 53). The exterior of this most utilitarian of WCs was curved and rounded, so that there were no corners

> where filth can collect, and there being no projections that can be kicked or knocked off, and used as missiles, the closet is specially adapted for positions where it is liable to rough, even violent treatment such as prisons, police courts, asylums, hospitals, schools etc., although it is equally suitable for general use where simplicity and cleanliness combined with strength are required.

It is important to realise how much of the pioneer work in the field of the developing WC was originated by individuals or small family concerns. Many of the Victorian plumbers were solid respectable men with solid respectable businesses. Above all they were practical men with the desire to get things done. It would be hard to find a better example than the firm of Humphersons in Chelsea, London.

The founder of the present firm, Edward Humpherson, had started his working life as a silversmith. About 1876 he diversified into plumbing, a decision which was to have far-reaching effects. Photographs show him as a splendid upright figure with a bushy moustache and, the inevitable symbol of success for a man of those days, a gold albert (watch and chain) draped across his broad chest. One of Edward Humpherson's maxims was that if one was to manufacture and sell an article it was necessary to understand not only how it worked but also how it was made and used. For this reason two of his sons, Frederick and Alfred, joined him after serving an apprenticeship with another famous Victorian plumber, Thomas Crapper. There is little doubt that they were the star pupils of their master. Their daily work was supplemented by intensive night-school tuition which covered a large range of subjects, like metallurgy, mineralogy, physics and chemistry, which could be expected to be of use to a practising master plumber.

Edward Humpherson's family consisted of five children, four boys and one girl. As already stated two of the sons were to join their father in business in Beaufort Street, Chelsea, whilst a third son moved to Newton Abbot in Devon, where he founded the firm of Hexter Humpherson which manufactured glazed pipes and other items of sanitary importance. But undoubtedly the most development was to take place at the Beaufort Works in Chelsea, and before long the firm proudly announced itself as 'Patentees and Manufacturers of Sanitary Appliances'.

Solid and hardworking as Edward Humpherson was, it was the original minds and enthusiasm of Frederick and Alfred which were responsible for the firm achieving the eminence that it did in plumbing matters. They produced a large range of products which were sent all over the British Isles, and used by other manufacturers. One of the earlier successes was an 'Improved Syphon Cistern' patented by Frederick in 1885. This remarkably efficient cistern received the only prize medal for flushing cisterns at the International Inventions Exhibition of that year. The prize was a handsome bronze medallion bearing the head of Queen Victoria on one side. Of the cistern, *Plumbing Equipment News* (February 1969) says that it 'may well have been the forerunner of a system which has not altered greatly throughout subsequent years'. Truly a remarkable tribute to a small family firm.

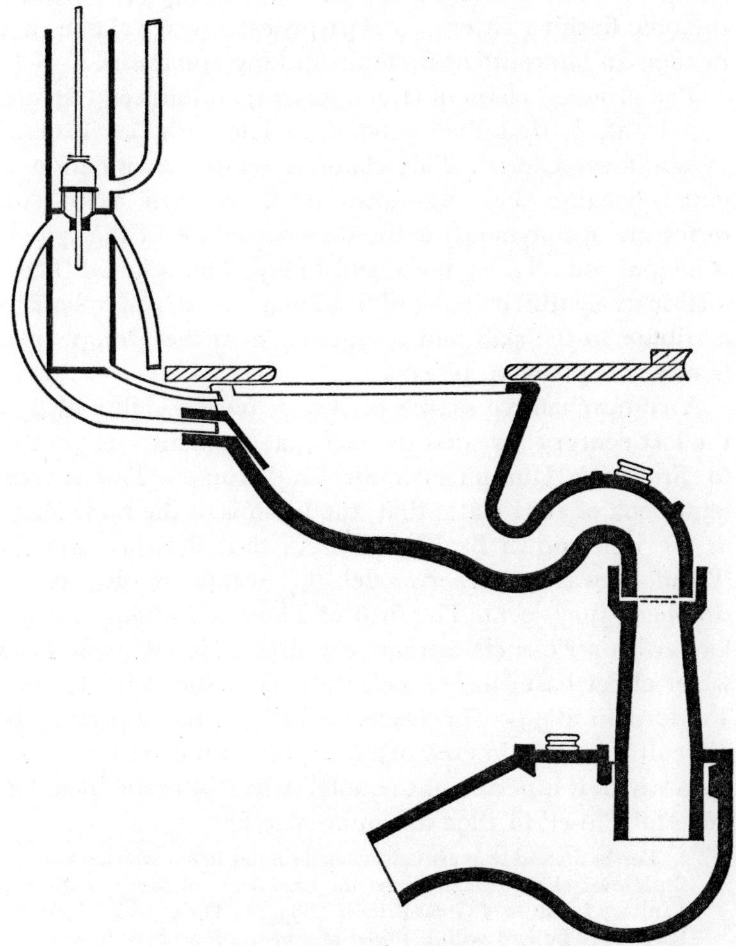

Fig 4 The early siphonic closet illustrated here was too complicated in construction. Later models were much simplified

Other inventions and patents were to follow. In 1886 there was a device for forming sockets on lead pipes, in 1888 a siphonic flushing cistern, in 1891 pedestal water closets, and in 1892 an intermittent siphonic flushing apparatus.

The proudest claim of Humphersons, and indeed the most important, is that they produced 'The Original Pedestal Wash-Down Closet'. This claim is worth examination in detail because the wash-down or flush-down closet (the terms are synonymous) is the direct ancestor of the type in principal use all over the world today. The original 'Beaufort' scarcely differs in detail from many modern appliances, a tribute to the skill and inventiveness of the Humpherson family (see plate 24, p 109).

A comprehensive search of British patent rights filed in the last century revealed the fact that a patent was granted to Frederick Humpherson for his 'Triune'. The current handbook of 1891 states that 'the basin is of the same shape as the well known Beaufort Closet', thus showing that the 'Beaufort' was an earlier model. But events are often not as simple as they seem. The firm of Thomas Twyford has put forward a serious claim that 'the first pedestal wash-down water-closet basin in one piece was introduced by Thomas Twyford in 1899—The Deluge'. This would appear to be difficult to sustain in view of a pamphlet still in existence and from which it is pertinent to quote. Advertising the 'Beaufort Pedestal Closet' in 1892 it is quite specific:

> The basin and trap are constructed similar to our well known flushdown closet which gained the Certificate of Merit of the Sanitary Institute of Great Britain, 1885. . . . The outside of the basin may be had with a raised ornamentation of ivy leaves, either plain or coloured. . . . All white basin and trap *in one piece* £1–16–0.

As if this was not enough to dispose of claims, other than Humphersons, to have produced the first wash-down closet,

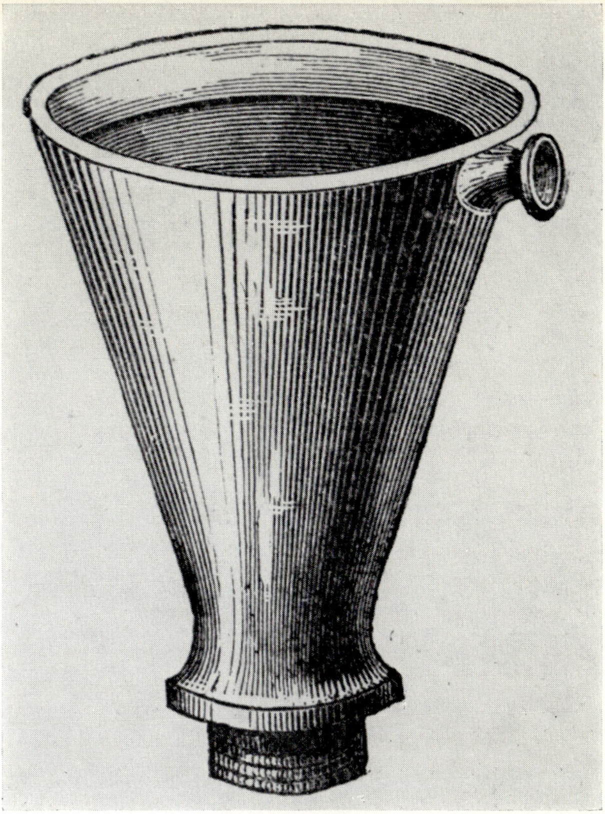

Plate 9 (*above*) The hopper closet had the advantage of being cheap. Hence it was considered highly suitable for the servant classes

Plate 10 (*below*) The 'Jar' made by Hellyers, was stated to be 'adapted for positions where it is liable to rough, even violent treatment'

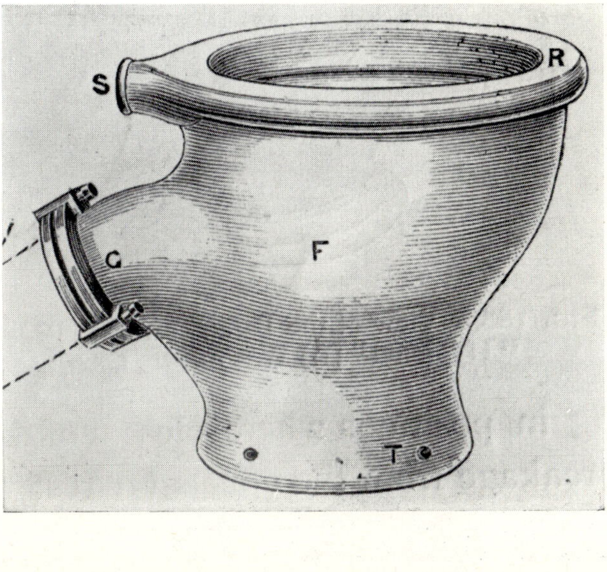

Plate 11 (*above*) The poor man's decorated closet. The dark rim reveals a cheap caneware basin and the later addition of a brick surround suggests it was not built to last. The tulip decoration is nevertheless ornate

Plate 12 (*below*) Though Hellyer detested the 'architectural' style he produced this model 'to meet the wishes of my friends for some little ornamentation'

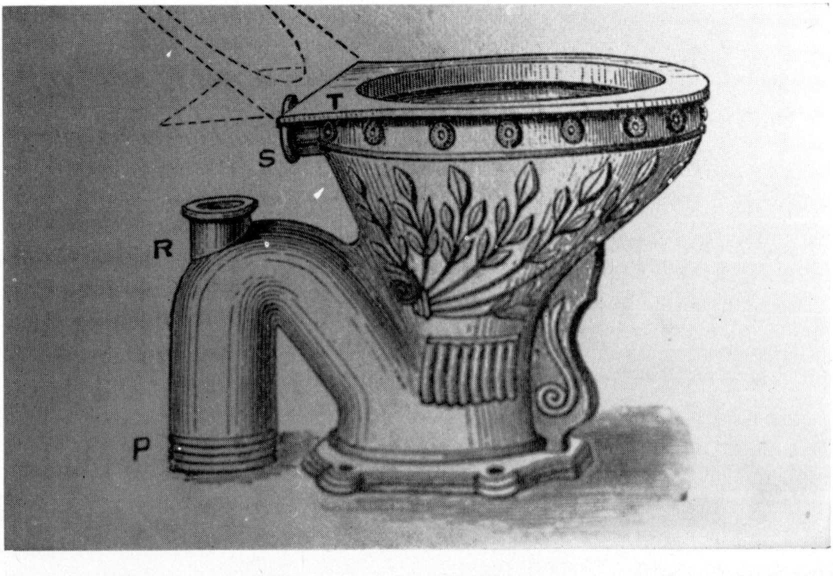

research was undertaken by Messrs Carpmael & Ransford, Chartered Patent Agents of Chancery Lane, London. In a letter dated 18 March 1970 they wrote:

> . . . we have conducted a subject matter search under the general heading water-closets and more particularly under the heading of pedestal water-closets of the washdown type, and from the result which the search revealed together with information gleaned from the publication . . . supplied, we have no reason to doubt the correctness of the statement 'The Original Pedestal Wash Down Closet' appearing on the advertisement entitled the 'Beaufort Pedestal Closet'.

Another claim by Lawrence Wright in his book *Clean and Decent* that the wash-down closet was invented by D. T. Bostel in 1889, will similarly carry no more weight than that of Twyfords, and Humphersons' evidence seems irrefutable.

It must not be supposed that at this period in their history Humphersons were only turning out their superb 'Beaufort' model. They had not abandoned the old valve closet which was still much used. They were able to supply the 'Perfect Elastic-Valve Closet', two of which were still working in London as late as 1970. It was claimed to be 'noiseless in action. . . . It is fitted with a large flushing rim basin, and an improved and ventilated overflow trap. The valve . . . can be easily renewed without the expense of taking the closet to pieces.' In best-quality china with a blue basin it cost £4 5s and in second-quality white china the delighted buyer could take it home for £3.

Whilst an efficient cistern and wc bowl are essential requisities for waterborne sanitation they must be combined with a reliable, non-porous drainage system. Victorian England had been startled out of its insanitary complacency by the *General Report on the Sanitary Conditions of the Labouring Classes of Great Britain* by Edwin Chadwick, mentioned in Chapter One. Unfortunately what was reported of the labouring classes in 1842 was true for all the big cities. Chadwick's report was followed by one by Dr Southwood Smith who reported on conditions in London with equally alarming results. This revelation disturbed Londoners and was to lead to the foundation of a new industry and a fortune for the Doulton family in Lambeth.

Chadwick had such faith in sanitation that he believed the demise of the medical profession was assured once the country had an efficient sanitation system. He pressed most assiduously for his beloved 'Sanitary Idea', hoping to impress upon officials and administration alike the importance of cleanliness and the means to attain it. He made the simple observation that epidemics were largely the prerogative of the poor who had to endure defective drains, excrement-contaminated situations and the like.

In 1846, four years after the Chadwick Report, young Henry Doulton, son of a potter, decided to manufacture stoneware pipes on his own initiative at a newly-opened factory at 63 High Street, Lambeth. His indebtedness to Chadwick was acknowledged in a speech he made many years later, in 1892, when he said:

> It was not until 1845 that any suggestion was made to employ pipes of pottery for sewerage purposes. The proposal emanated from the late Edwin Chadwick who before he died earned the title of 'Father of Sanitary Science'. Forseeing the magnitude of the demand likely to arise, I did not think it advisable to graft the business to the existing factory, but commenced special works in High Street, Lambeth, devoted entirely to the production of pipes for town and house drainage.

With initial capital of £600 he began the difficult task of manufacturing stoneware pipes. Many leading sanitarians were unenthusiastic about the need for such pipes and the money was soon swallowed up. A reluctant loan of a further £400 from his father helped to keep him solvent. Soon the experiments which he carried out on drainage and stoneware pipes started to convince sanitary engineers of the suitability of the material for the job. The old brick conduit, once regarded as adequate, was finished and in its place came glazed pipes of a hitherto unattained hygienic standard. In 1848 the Public Health Act had been passed in the British Parliament and it gave great impetus to the increasing use of stoneware products. Henry Doulton was later to write:

> Glazed stoneware pipes were first used by Mr John Roe, of the Holborn and Finsbury Vestry and by Mr J. Phillips, of the Westminster Commission of sewers. . . . The first great impulse to sewer and drainpipe making was given by the Public Health Act in 1848, when works of town sewerage were commenced in various parts of England.

By 1847 Doulton had found it necessary to expand production and factories were opened in St Helens and in Dudley. It was a natural progression from stoneware pipes to WC bowls and before long the firm was supplying a rapidly expanding market with these much improved articles.

In 1899 they were advertising themselves as Doulton & Co of Lambeth, Paisly and Paris. Among their appliances was the 'Patent Syphonic Closet' made in one piece earthware with a range of variously decorated bowls. Complete with cistern and full equipment this model cost £6 11s in Blue Magnolia and £6 15s 6d in Wild Rose decoration. Whatever may have been the later successes of the firm there is no doubt that their singular contribution to the development of the WC was their early work on the humble glazed pipe.

The combination of the plumber and the potter yielded rich dividends for a few fortunate companies and big empires have grown from small beginnings. Typical of such growth has been the firm of Twyfords, now a huge concern whose products are sent all over the world. Yet the start of this mighty enterprise was humble enough and hinged on the genius of one man, the original Thomas Twyford, whose skill at burning clay was to make possible new and improved articles of sanitary ware. With reference to him it has been said, 'it is not alone in the mixing of clays, the forming of vast pieces of ware and the glazing thereof in which success consists, but in the burning'. It is proposed in a subsequent chapter to deal with the process of turning clays and earths into receptacles and the name of Twyford will receive its due acknowledgement. The ability of Twyfords to produce large items of sanitary ware in one piece was to revolutionise the industry.

The firm was founded by Josiah Twyford whose skill had been learnt from two seventeenth-century potters from Amsterdam, brothers by the name of Elers. The pottery continued quietly enough in Hanley but when his grandson Joseph William succeeded in 1872 great changes were to take place.

Until 1872 Twyfords had produced an unexciting if useful range of domestic goods and their share in the now expanding market for indoor sanitation had been modest enough. Joseph Hatton, in a history of Twyfords written about 1905, says:

> Mr Thomas William Twyford saw that there were possibilities of great development in the art and business of the sanitary potter who at that time was content with few exceptions to receive, for his share of a miscalled 'Sanitary Appliance' 2s for the earthenware, the brass and iron-founder getting from 20s to 50s. He thought this was hardly equitable in principle, and so he conceived the idea of the 'National' basin, made all in earthenware, to displace the old 'pan' closet and similar unsanitary systems. At first his innovation met with scant success, owing to the difficulty with which old-fashioned notions and customs are superseded. During the first year of its introduction not more than fifty were sold; the next year about 200; a year or two later, when its advantages from a sanitary point of view,

> were generally discovered and acknowledged, the sales reached about 10,000 a year.
>
> In the summer of 1885, when an enquiry was received . . . from a French architect of advanced views in sanitary engineering, who desired to know if he could be supplied with a basin to be fixed open and exposed, without any wooden enclosure, Mr Twyford replied in the affirmative, and had at once a design prepared and submitted to the architect, who was so pleased with it that 700 were at once ordered, to be fixed in some large residential flats . . . in Paris.

According to Hatton, this WC basin, which was sold under the name 'Unitas', was the 'pioneer of Pedestal closets', and was of the wash-out kind. It was a big improvement on earlier WCs as all external surfaces were exposed and could thus be kept clean.

In 1886 the 'Unitas' was exhibited at the Health Exhibition in London and was the subject of much controversy. There is some doubt whether this new type of WC would have received the early recognition it did, had it not been for the endorsement of the medical profession who in those days took a lively interest in such matters.

Despite its many advantages the 'Unitas' proved a difficult piece of pottery to produce and in its early days of manufacture, when potters were paid on a piece-rate system, it was hard to find men to undertake the work. Eventually Twyford was able to persuade some of his more highly skilled workmen to embark upon this new venture with results that were beneficial to employer and employee alike.

In 1889 Twyfords introduced their wash-down closet, the 'Deluge', and in 1894 the 'Twycliffe', which worked on a siphon system.

Contemporary catalogues provide a rich source of information about these masterpieces. In 1894, the year of its introduction, the 'Twycliffe' was advertised complete, including a porcelain basin and slop top, matching paper box in Corinthian pattern, mahogany seat, brass flush pipe and pendant pull for £14 12s. By 1901 the price had risen to £18—inflation has been ever with us. The 'Deluge', in Venetian pattern, came out slightly cheaper. In 1894 it cost complete £10 14s.

By 1899 100,000 of Twyfords 'National' closets were in regular use and the latest inducement stated:

> An Important Alteration has been made in the Flushing Arrangement of this Closet. The SUPPLY is now placed opposite the OUTLET, and a NEW REGISTERED FAN attached which makes the Closet more effective.
>
> The OUTLET is placed at the side in such a manner that the great and offensive objection of looking into the soil pipe is avoided.

Other pioneers in the development of the WC include the firm of Shanks, of Barrhead, Glasgow. Though remote from the main centres of activity in London and the Potteries they were not to be left out of the race. In 1896 they produced a lavishly illustrated catalogue modestly professing to be the 'epitome of Patented Sanitary Appliances embracing all the latest Sanitary Novelties up to date'. Of note was the 'Barrhead Siphonic Closet' in which Shanks claim to have solved the problems of this type of WC—with an 8in deep water seal and a surface area of water 13½ × 11in, 'the largest of any closet'. The description of this mighty machine grows lyrical:

> The siphonage is produced by the direct action of water from the service pipe, and notwithstanding the great depth of the seal the contents of the basin are discharged with the most powerful suction and the basin is filled again with clean water from a two gallon flush.

Many sanitary engineers wrote in glowing terms of this WC. One firm is on record as saying: 'We venture to congratulate you. . . . The marvellous work effected by a two gallon flush almost surpasses belief, . . . it far exceeds any other closets we have seen.' A Mr Kirtley RP, of Sunderland, wrote in 1895: 'I regard this, the latest development of closets, as the perfection of sanitary work.'

The Leverne cistern which Shanks produced as an accessory to their 'Barrhead Siphonic Closet' worked quite well on a 2 gallon flush. For an extra £2 this could be fitted with 'Dicksee's Patent Automatic Flushing Apparatus' and this was recommended wherever possible. Once Dicksee's patent was fitted the users could be assured of a quite remarkable chain of events when slops were thrown into it. Then siphonic action was started, but simultaneously siphonic action of the cistern started, 'so that slops are not only discharged from the basin but the basin is flushed and filled with clean water'.

Meanwhile at Hanley in Staffordshire T. C. Brown-Westhead, Moore & Co Ltd, with their eyes fixed on the export market, were exhibiting in Paris. In 1889 they won a Grand Prix and produced a catalogue in both English and French. Thus they could advertise a WC 'avec porte-siege brevete' (a seat extension), and produce such models as the 'Paris', the 'Hauteville' and the 'Parfait'. Yet despite the catalogue's French text, the good old English abbreviation 'WC' was as familiar on the Continent as it was at home, and it rapidly became an accepted form of lingua franca all over the world.

In the USA after 1880 there were many ingenious developments of the WC. In 1882 James E. Boyle came up with the idea of a pneumatic closet, but this was a very complicated device which was never successful. The following year J. Pickering Putnam produced his 'Sanitas Closet'. Its special advantage was extreme quietness in action, but again, the mechanism never caught on.

It would be impossible to list all the models manufactured

by pioneer firms in the history of the water closet. By the turn of the century it had become a veritable torrent. Suddenly the wc had become the Victorian status symbol, and as its popularity spread to the lower-income groups, firms were forced into producing less expensive models. Thus Dent & Hellyers turned out the 'Artisan', aimed at the working classes. This name proved so unpopular that it had to be changed to the 'Citizen'.

In the weaving sheds and cotton mills of the rugged North of England, the provision of sanitary services had to be cheap and reliable. The demand in this market brought the firm of Ducketts of Burnley, Lancashire, into prominence. Ever alive to the needs of the mill owners for their employees, they catered for what might be called the bottom end of the market.

The firm had been established by James Duckett in 1859. Like many sanitarians before him he had started life in a different discipline. A Preston builder of some merit, who undertook the erection of the original Ha'penny toll bridge over the Ribble, he turned his attention to the building of St Mary's Catholic Church in Burnley. During his train journeys between Preston and Burnley he noticed that in one locality the railway cutting had been driven through shale suitable for brick making. As a result he decided to build a brick works nearby and this later diversified into the production of salt-glazed pipes, no doubt following the example of Doulton before him. Shortly after the turn of the century, and as a result of experience gained in the manufacture of pipes, Ducketts began the production of sanitary fittings in white vitreous enamel, amber and cane. The whole range included urinal slabs and stalls, closet pans, wash-basins, sinks and channel pipes, which were sold all over Britain, in Europe and in South America. The firm also gained the distinction of having its urinals installed on Berlin's famous Unter den Linden.

The first closets made by Ducketts were trough closets, as noted in the chapter on 'Seats and Cisterns'. This was a revolting piece of apparatus, rather like a horse trough, the bottom of which was covered with water. In its most spartan form the trough was edged with a hardwood rail and the performer must have felt he was sitting on a branch overhanging a polluted stream. Fixed or hinged ring seats could be obtained as optional extras. In 1913 a 12ft range for four persons cost £7, whilst the model for eight people was £10 and was 24ft long.

Shortly after the introduction of trough closets the firm began the manufacture of individual tipper closets which were to replace the old insanitary pail closets so common in the industrial North and Midlands. These are considered elsewhere.

James Duckett was followed by his son Alfred, who in his turn was succeeded by his son George. The firm was later sold to Temperley & Sons of Todmorden who finally sold out to the Hepworth Group, one of the biggest, if not the biggest pipe and conduit makers in both the United Kingdom and the United States.

Although the firm no longer exists in its own right the name is still to be found on thousands of tipper closets still working as efficiently as the day they were first installed.

There is no doubt that in the nineteenth century Great Britain led the world in the production of sanitary ware. Initially the Americans preferred to import their supplies from Europe and particularly England, but it must not be supposed that the Americans had been idle and Europe's early lead in sanitary development was not to last for long. Before the end of the nineteenth century the USA was manufacturing sanitary ware as good as any that came from abroad. Once the reluctance of the Americans to buy their own products was overcome they began to swamp the American market.

The American sanitary ware industry had started inauspiciously. In 1847 Thomas Maddock, who had worked as a pottery decorator at Longport, Staffordshire, came out to New York with his friend, William Leigh. The two men started work by decorating china in Spruce Street, New York, and it is reported that in 1853 one of their dinner services reached the White House for President Pierce's use.

By 1873 Maddock was bold enough to strike out in a new venture, forming a partnership under the name of Millington, Astbury & Maddock. Their factory in Trenton manufactured earthenware for which there was now an expanding market. It was about this time that Maddock conceived the idea of producing sanitary ware for the American market. Like pioneers before him in England he soon came to realise the difficulties of the task he had set himself. Previously Isaac Davis of the American Crockery Company in Trenton and James Carr in New York had tried and finally abandoned their attempts.

Maddock was made of sterner stuff. He soon discovered that one of the basic problems he had to solve was to determine the correct proportions in the ingredients for making large pieces of earthenware such as he envisaged and which he saw coming in shipload after shipload from England. He too was to find, like Twyford, that one of the great arts of the potter was in the firing. Doggedly, obstinately he struggled on to the dismay of his partners who could only see disaster in his work. The story reads so much like that of Twyford's. There was trouble with the workmen whose piece-rate system had to be changed to a day-rate one. There was a high percentage of failures in the firing; dunting and crazing were commonplace. Even the acceptable finished articles were not of a very high quality and the task of selling to the American agents, used to much superior ware from Great Britain, was perhaps the hardest task that Maddock had to undertake.

In 1873 four British firms—Brown-Westhead & Moore Ltd, Twyfords, Dimock, and Brown of Paisley—were exporting their products to the United States. This represented stiff opposition for Maddock's ware. Eventually towards the end of 1873 he succeeded in placing a small order with Miller & Coats of New York. On this historic occasion the American sanitary industry may be said to have been born. Yet most Americans who bought their own native products were unaware of the truth. In order to induce the good American citizen to part with his money each basin was carefully stamped with a lion and unicorn and the magic formula 'Best Stafford Earthenware made for the American Market'.

It was not long before Maddock was able to supply them as genuine American products without the need for an English pedigree. He went from success to success adding the firm of Waeffelaer & Duysters to his list of buyers and soon his factory, working to capacity, was seeing its products in use all over the United States and in the hands of countless suppliers.

The field for sanitary ware was suddenly wide open and inevitably other firms began to compete with him. A list compiled in 1909 gives some idea of how quickly the industry expanded:

Thomas Maddock's Sons Co	1873
Taylor & Houdayer	
Enterprise Pottery Co	1879

Isaac Davis	1883
Trenton Pottery Works	
Fell & Throp Co	
Crescent, Trenton	
Delaware Pottery Co	1884
Equitable Pottery Co	1888
The Great Western Pottery Co	1889
Hammill & Brown	1890
The Willets Manufacturing Co	1909

CHAPTER FIVE

Materials and Manufacture

In the early days of the manufacture of wc basins, one of the many problems was to find a material which would be tough, impermeable to liquids and reasonably easy to work. Harington had suggested a basin of brick, stone or lead dressed with pitch, rosin or wax. Any of these combinations must have been full of shortcomings. Later attempts were made to utilise marble but it was difficult to carve, expensive and far from satisfactory. Since those pioneering days new materials have emerged, including stoneware, earthenware, fireclay and vitreous enamelled porcelain.

One of the early breakthroughs came with the discovery of salt glazing. By this process easily worked materials could be covered with an impervious glaze which offered a new resistance to stain and liquid. As long as the articles were not complex in shape and thus difficult to fire then earthenware could be used and salt glazed. As has been seen, it was this simple fact that permitted the manufacture of the salt-glazed pipes which Doulton popularised and which revolutionised sewage disposal.

Much of the early work in the manufacture of sanitary ware took place in the Staffordshire 'Potteries' district of England where a large labour force with the necessary expertise was available. Salt glazing has a long ancestry and its origin was described in the *Pictorial Gallery of Arts*, published in 1885:

> According to the current story, at Stanley Farm, situated not far from Burslem, a servant was boiling in an earthenware vessel a strong lye of common salt, to be used in curing meat. During her temporary absence the liquid boiled over, and some ran down the sides of the vessel, covering the surface with a liquid which on cooling appeared as a glaze. Mr Plummer, a potter of the neighbourhood, being acquainted with the fact, speedily made profitable use of it, and established the manufacture of common brown glazed ware.

Plummer's discovery was further elaborated by the Elers brothers of Bradwell in the last quarter of the seventeenth century. At a certain stage in the firing shovelfuls of wet salt were thrown into the kilns and a tough impervious glaze imparted to the vessels therein. Articles made of such materials were of limited use in sanitary circles where the requirements were exacting. Soon after the beginning of the eighteenth century, whiter and hence more pleasing clays from Dorset and Devon were being used by the Staffordshire potters for coating the vessels, thus disguising the rough clay base. Then about 1700 a considerable improvement took place when William Astbury introduced calcined flint stone, initially as a surface wash on articles. Later it was incorporated with other clays as a basic raw material. An account of this important discovery is found in *Twyfords: A Chapter in the History of Pottery*, by Joseph Hatton:

> . . . pottery in general is indebted to Astbury for another exten-

> sion in the art through an incident which came under his observation during a journey to London in 1700. He had arrived at Dunstable, when his horse seemed to be rapidly going blind. At the tavern where he stayed the hostler undertook to doctor it. Taking a flint stone, he burnt it until it was at red heat. Then he pulverised it very fine, and by blowing a little of the dust into each eye occasioned a discharge of matter that greatly relieved the suffering animal. Mr Astbury was struck with the whiteness of the calcined flint, the ease with which it was reduced to powder, and its clayey nature when it was discharged from the horse's eyes; and it at once occurred to him that the material might be usefully employed to give a different colour to the pottery he was producing. Returning home, he made a successful experiment with the calcined flint; and, using it in his clay, obtained thereby a preference for his wares which brought him considerable advantages in the way of profit and reputation.

Before considering the way by which large articles of sanitary ware such as we know today are made, it is pertinent to examine the different materials available. One of the early ones, used extensively for the simpler closets like the hoppers, was stoneware. Strictly speaking this is a pottery, the main characteristic of which is an opaque vitrified body which is practically impervious. The term is, however, loosely used. In earlier days when the manufacturers of stoneware were unable to keep up with demand sanitary articles were produced in cheap fireclay, which was then glazed to resemble stoneware. This was done by interposing an intermediate slip, or engobe, between the fireclay and the glaze. This engobe, consisting of clays which would burn white, was compounded with felspar to make it stick. Thus fireclay articles could be passed on as stoneware to the unsuspecting customer.

Another variation, this time of stoneware itself, was cane-ware to which many nineteenth-century advertisements for sanitary ware refer. After the first firing was completed a composite glaze of Cornish stone, felspar, whiting and flint, in the proportions of 3:1:1:1 respectively, was fused to the basin and produced its own peculiar colour, a shade of yellow.

A much-used material which has survived into the twentieth century is glazed earthenware. Technically this can be subdivided into two kinds: coarse and fine. The name covers a wide range of different kinds of pottery which find many uses quite apart from sanitary ware. All of them are characterised by a porous body, or 'biscuit', which will be described later. Such materials require a glaze to cover this porous body.

The coarse earthenwares are often made of a single clay and in the early days of pottery manufacture in Staffordshire these clays were obtained locally. Occasionally two or more clays were combined together with sand or broken pottery mixed in. After firing a glaze was added, but such materials were of use only for the cheapest sanitary ware.

Fine earthenware, on the other hand, is a carefully compounded mixture of fine clays and flints skilfully blended and then glazed. Such material was widely used by British potters of the nineteenth century once the initial difficulties had been overcome. Still widely used today it has a serious rival in fireclay.

Fireclay, unlike earthenware, is a single material which has the advantage that it can be fired and glazed in one process. A natural silicate of aluminium, it is generally found in connection with coal seams. It has proved a very suitable material for sanitary ware and is increasingly used by modern manufacturers. It has an impervious body and cracking of the glaze is not as serious a disadvantage as it would be in earthenware where a porous body exists. Fireclay is being used increasingly in better-class work and is generally preferred in Europe.

Plate 13 (*above*) The rich man's closet. An expensive closet embossed and lined with gold. It shows that decorations were often applied to the inside, outside and on the rim as well

Plate 14 (*below*) Chrysanthemums were perhaps the commonest flowers to be found adorning the wc basins of discerning Victorians

Plate 15 *(above)* A favourite pattern for valve closets was the richly ornamented 'Panorama' which sold in thousands

Plate 16 *(below)* This still life of fruits and flowers shows the degree of decoration often applied to early basins. The valve is clearly seen in the base

Finally, mention should be made of vitreous china, a material much used in the United States. This material, somewhat resembling the fine earthenware of England, is composed of a mixture of clays which has to be fired first and then glazed. It is said to have a less porous body than English earthenware and is thus less absorbent should the glaze become cracked. In the beginning American manufacturers had sought to emulate the English potters by importing their materials from abroad, but this was a costly business soon discontinued when native raw materials proved more than adequate for their purpose.

There was a great step forward in 1877 when Charles Harrison, working in New York and perhaps fired by Maddock's success, patented a bowl and drip tray in one piece for use with a pan closet. In this product Harrison showed that the problem of firing a bowl and surmounting drip tray (rather like a table top) in one piece had been overcome. At last it was seen that an all-porcelain WC was a possibility and lead and wood could be done away with.

Today some of the old processes have been automated but the production of sanitary ware remains a highly skilled operation. It might seem at first sight an easy enough undertaking—thumping wet clay into a mould and then firing it in a kiln until it is baked hard. Nothing could be further from the truth as the production of the humble WC bowl is a masterpiece of manufacture. Hatton, in 1905, gave some idea of the processes involved in making such articles in fine earthenware. Though dealing with work done at Twyfords, he underlined the problems which all manufacturers of those days had to face.

As already noted, fine earthenware is blended from different materials, each one carefully chosen to give the finished article certain properties. Basically there are four ingredients. Blue or ball clay is found usually in tertiary deposits in Devon or Dorset and provides the necessary body for the finished product. This is combined with Cornish stone, a derivative of granite, which is rich in felspathic material and acts as a flux. To these are added flints which occur in chalk strata and are almost pure silica. Some of the best are to be found in the neighbourhood of Dieppe, in northern France. Finally china clay, from Cornwall, adds whiteness to the finished pot when it is fired and is an essential ingredient of fine earthenware.

The Staffordshire Potteries had originated because of the presence of coal and usable clays. With diversification into substances like earthenware, Twyfords, along with others, were faced with the problem of transporting materials from relatively distant places. By 1905 the canal link between the rivers Mersey and Trent had come into being. Thus stones and flints could come up the Mersey to Runcorn in Cheshire, be off-loaded into barges and travel via the Bridgewater Canal to the works at Cliffe Vale. On arrival the materials had to undergo various processes before being blended together and passed to the potters. The description of this leads one into an esoteric world of slip and slop, pugging and pressing, glosts and saggers, badgers and blungers. . . .

The basic clays, china and blue ball, require little treatment as they contain few impurities and are fine enough not to need further breaking down. Hence they are agitated in water until the particles are in suspension, a process known as 'blunging'.

On the other hand the Cornish stone and flints have to be reduced to fine particles before they can be blended with the clays. As Astbury found out in 1700, flints have to be calcined before they can be ground to fine powder in a mill.

This consists of heating them to redness in a kiln and then allowing them to cool. In Astbury's days the demands had been met by the simple process of pounding them in an iron mortar until 'perfectly levigated'. This proved a highly dangerous occupation for those engaged in it and their expectancy of life was dramatically reduced. By 1713 a plumber and glazier called Thomas Benson or Bedson (different versions of his name exist), of Newcastle-under-Lyme, conceived the idea of crushing the flints in water, thus stopping the dust which did so much harm to the workmen from arising. In 1713 Benson was granted letters patent and an account of this is given in the *Staffordshire Advertiser* of May 1833:

> . . . that the petitioner has, with great pains and expence, invented and brought to perfection an Engine, or new method for the more expeditious working the said Flint Stone, whereby all the said hazards and inconveniences attending the same, will effectually be prevented; and that he proposes to perform the same in manner hereinafter described viz.: the Flint Stones are first sprinkled with water, insomuch that no dust can rise to the hurt or danger of the workmen; then crushed as small as sand, by two large wheels of the bigness and shape of mill-stones, made of iron, to turn round upon the edges by the power of a water-wheel, and afterwards conveyed into large pans made of iron, . . . in which there are large circular balls, which by the power of the water-wheel above mentioned, are driven round by such swiftness of motion, that in a little time the Flint Stones so broken are made as fine as oil itself, without the use of lawns.

About 1758 James Brindley, a millwright, built a windmill near Burslem specifically for the purpose of grinding calcined flints. Later watermills were pressed into service and eventually steam-powered mills came into operation. The mill at Twyfords described by Hatton was worked by a 'large horizontal compound condensing engine, of 350 horse-power'. In such a mill both the calcined flints and the Cornish stone were ground so that the particles were fine enough to remain in suspension in water. Only then were they ready to be mixed with the blue ball and china clays.

All four materials, now designated slop flint, slop stone, china slip and ball slip were mixed together with water in a blunger until the whole was a homogeneous mass, referred to as slip or slurry. A typical mixture for sanitary ware would consist of about 46 per cent ball clay, 23 per cent china clay, 21 per cent flints and 10 per cent Cornish stone. Such slip was pumped through a series of lawns which consisted of tiers of wooden trays whose bottoms were covered with silk or fine phosphor-bronze wire mesh. These lawns were necessary to remove any large pieces of raw material which may have remained in the slip.

The next stage in the many processes needed was to pass the slip over electro-magnets to remove any particles of iron which would produce an unpleasant speckled appearance to the finished basin were they left in.

Then excess water had to be removed in a type of press called a finishing ark where it received its final mixing before going to the filter presses. From here the clay went into a 'pugger', a machine which compressed it and ejected it from a square nozzle in a putty-like consistency. Then at last the potter could lay his hands upon it and begin the lengthy and arduous task of converting it into a WC basin, a lavatory bowl or a bidet.

Essentially the process of manufacturing sanitary ware is one of introducing the putty-like clay into plaster moulds, a task known as pressing, from which they are removed later for firing. Today injection methods have removed some of the difficulties of pressing but in Hatton's time it was a painstaking and highly skilled process. One of the difficulties was to ensure that no air spaces were left in the mould which, expanding in the heating in the kilns, would lead to cracking of the finished piece.

The moulded articles were then dried in ovens until enough water had been driven off for them to sustain their own shape. The moulds were then removed and, provided they were in good condition, could be used again. The ware itself underwent a sprucing-up process, being smoothed down with sponges, knives and horn instruments, until the surfaces were without blemishes. Then came a long slow drying to get rid of all the water and at last the firing in a furnace which often took the best part of a week. At the end of firing, just as today, the articles had a porous body and a rough surface. This rough unsuitable material, known as bisque or biscuit ware, had then to be covered with a layer of impervious glaze. In the old days when elaborate patterns were much in vogue they were applied to the basins before glazing took place. Many of these WC bowls are now collectors' items, and the method by which these decorations were applied is interesting. If the ware was to be printed (ie receive a pattern), it was sent to the printing-shop, where the partly baked pieces received their underglaze decorations in one or more colours. Hatton gives a good description of this work in his book:

> Printing is done by transfer, the design being first engraved on a copper plate. To obtain the transfer, the printer makes up a mixture of colour (compounded from metallic oxides etc.) with specially prepared oil. This is spread evenly over an engraved copper plate kept warm on a stove, and the superfluous colour removed by scraping and 'bossing' until there remains on the plate only that which is retained in the sunk or engraved portion. Specially made printers' tissue paper—previously cut to size and damped with size-water—is then carefully placed on the plate and passed through a press (which consists of two rollers wrapped with several thicknesses of flannel) thus receiving an impression of the design in printing colours from the

> copper plate. This is handed to the women (called transferrers) who first cut off with scissors any superfluous paper, then carefully place the transfer (with the colour side next the ware) on the article to be printed in the desired position, and finally well rub the same on with flannel rubbers. The ware, being absorbent, retains the oil and colour together with the paper, but the later is removed by sponging or washing with water. This, strange to say, can be done without in any way injuring the transferred design. The ware is then ready for 'hardening on'. If the article is to be enamelled 'under-glaze', it is, after being printed, sent up to the Painting or Decorating shop to be painted. This consists in embellishing in various colours by aid of a camel hair pencil the design to be printed on the article, or in the free painting in colours, without any print, of a decorative design on the face of the absorbent ware.
>
> The printed or painted article has now to be passed through a kiln known as the 'hardening on' kiln. This has the effect of removing the absorbed oil, leaving the colour on the ware. The reason for this is that oily or greasy surfaces will not take glaze. After being passed through the kiln the colours, being deprived of the oil, could with some little effort be rubbed off the surface of the ware; but the operation of 'dipping', for which it is now prepared, prevents this possibility and makes the ornamentation practically part of the ware.

The last stage of all comes with the application of the glaze, which covers the entire surface of the bowl with a thin impervious layer of glass beneath which any pattern is imprisoned for ever. Originally most glazes for sanitary ware contained lead, and these were a great danger to the workmen engaged in the dipping house, where articles were treated prior to the final firing. Later glazes were developed which were free of lead and one not infrequently encounters WC bowls and urinals which proudly announce this fact.

Thus it can be seen that the production of fine earthenware basins is a complex task for many skilled hands. Just as Hatton has written his account of Twyford's undertakings so Darwin, in his *Botanical Garden*, summed up the work succinctly on behalf of another firm:

Gnomes as you now dissect with hammer fine
The granite rock, the noduled flint calcine;
Grind with strong arm and circling chertz betwixt.
Your pure kaolins, and petuntses mixt;
O'er each red saggar's burning cave preside,
The keen ey'd fire-nymphs, blazing by your side,
And pleas'd, on Wedgwood may your partial smile,
A new Etruria decks Britannia's isle.

CHAPTER SIX

Decorations and Nomenclature

QUITE apart from technical developments both in cistern design and in the bowls themselves, the inventive genius and artistic skill of the Victorians burgeoned into full flower when it came to the question of decoration. In a sense, today we are passing through a much more inhibited age in much of our household furnishing. The Victorians followed entirely different conventions and gave full range to their fertile imaginations.

Initially, when the bowl had been camouflaged by its wooden surround, any decorations applied to it were confined to its interior where they could be seen. With the decline of the valve closet, and the consequent disappearance of the wooden wall, decorations to the WC bowl received a new lease of life. With the wash-out and wash-down models which stood naked and exposed to all eyes, it became possible at last to regard the WC bowl not only as a functional article but also as an artistic one. The outside of the bowl could be embossed or it could be coloured.

There were obvious limitations in the three-dimensional style on an article where hygiene was a necessary consideration. The chubby cherubs and cornucopia so beloved in sculpture of the period were thus highly unsuitable. Nevertheless, embossed or 'architectural' styles were quite common and ranged from simple fluting on the pedestal to more elaborate patterns bearing fanciful names. That all sanitarians were not in favour of such treatment to the basin was evidenced by Hellyer's remarks on the subject:

> . . . and so elaborate and so heavy is the external ornamentation upon some of their exterior parts, that a humane man hesitates to use them for one of the purposes for which they are fixed, fearing to give the chambermaid a life-long labour in making them clean and wholesome again; for, with the utmost care in using such closets as urinals, it is impossible at all times to urinate into the basin so neatly that splashes, mishaps and drops . . . shall not fall upon the edge of the basin and run down over its exterior parts.
>
> The heavy ornamentation of ceiling centre-pieces with which the jerry builder often seeks to captivate his purchasers are very simple compared with the very heavy ornamentation that the potters are now putting upon their closets which are to stand without enclosures. Floral and architectural patterns in deep relief cover the exterior of many of their closets; even recumbent lions, and swans in full sail, with closet basins on their backs, and other incongruous combinations are being produced, . . .
>
> To meet the wishes of some of my friends, for some little ornamentation upon the Pedestal Hygienic, for fixing in the best wcs, it has been decorated with a slightly raised pattern. . . . [See plate 12, p 55.]
>
> When so desired, the closet can be had in colours—in Indian Pearl-leaf pattern, picked out in gold, or it can be marbled inside and out, to suit different tastes.

The popularity of embossed designing grew out of development of pedestal models. Once the closet was able to shed its wooden surrounds it seemed that the Victorians felt it must be made more beautiful, it must somehow give the appearance of not looking what it was, but must pretend to be a sculptured bowl or a Corinthian column or something equally exotic. This was carried to extremes in the 'Lion' and the 'Dolphin' of Messrs Armitage, described later.

For the main part, however, embossed styles were relatively restrained and designers were well aware of the potters' difficulties in producing these complex articles. Often designs were applied to the unfired basins in the form of strips of different coloured clays, referred to as 'Pâte Duré', which fused with the rest of the material during the process of firing.

Undoubtedly the most popular form of decoration was by means of underglaze patterns, the technique of which is described in another chapter. It was a process that gave free rein to the fertile artistic imagination, was relatively easy to carry out, and once completed the pattern was fixed—immutable and hygienic.

There were occasions when it was found necessary to apply decorations after the glazing was carried out, in particular with fireclay and similar materials, where the firing and glazing are done in one process. In such cases hand painting was used and patterns of birds, fruit and flowers were popular. Such designs did not have the permanence of underglaze decorations and generally speaking fireclay objects have received less ornate treatment than earthenware. It is partly the increasing use of fireclay that has led to a decline in coloured decorations.

Another method, known technically as ground laying, was to cover the glazed basin with a thin coat of colour and subject it to a third firing. This had the disadvantage of increasing production costs and was restricted to the more expensive items of sanitary ware.

But perhaps the most expensive form of decoration was that of gilding. Specially prepared gold, ground down with alloys and flux, and made up in a turpentine and oil base, was applied by brush on to an already embossed pattern. This was then fired again in the kiln and was finally polished by hand to give it a bright lustrous appearance. Gilding was perhaps the ultimate luxury in so mundane an object as the WC bowl, but some rich Victorians could so indulge their fancy (see plate 13, p 71).

Perhaps one of the most thoughtful approaches to the large subject of decoration came from the firm of Shanks of Barrhead, Glasgow. By 1899 a great deal of expertise had been acquired and they had turned their attention to the whole of the bathroom, with matching WC, bath, lavatory basin and bidet. The builder of a new house at the turn of the century could pick from a lavish catalogue a tasteful 'Ribbon and Shell' design, or 'Renaissance', 'Empire', 'Gold Line', 'Touraine', or 'Baronial'. There were also raised decorations in white, or white with the decorations picked out in a soft blue, rather like Wedgwood in colour.

Additionally there were the ever-popular flower patterns. 'Keele' consisted of large red and blue flowers with trailing leaves and 'Orchid' had big blue flowers. 'Carnation', 'Morning Glory', 'Chrysanthemum', 'Hydrangea', and designs named 'Chantry' and 'Wolsley' all sported flowers as the basis of their decoration (see plates 11 and 14, pp 54 and 71). And if the jaded eye was still not satisfied there were variations of most of these under the generic term of 'Mulberry', generally in tasteful grey and blue shades.

There seemed to be no end to the flowers. The Garden of Eden could not have bloomed more profusely. Whilst chrysanthemums were always a firm favourite among Victorian families the more adventurous searcher could find honeysuckle, peach, poppy, thistle, roses, daisies and no doubt others to bring colour into his WC.

Even these did not exhaust the offerings put forward to tempt one. There were many non-floral patterns, some with fanciful names. 'Pink Lucknow', 'Brown York', 'Olive Green Chicago', 'Grey Plassy', 'Mulberry Marble', 'Cretonne', 'Mazarine Blue and Gold', 'Chocolate Galatz', 'Dacca Mulberry', 'Pink Teviot', 'Dark Blue Leghorn' and 'Decor Louis XV' all existed, though few have survived to the present day.

Even more elaborate patterns were to be found. Particularly popular was an ornate garden scene, called 'Panorama', with Italianate overtones (see plate 15, p 72). This, like the ever popular Willow Pattern for dinner services, was turned out by many different firms in slightly different versions. Bowls of fruit, baskets of flowers (see plate 16, p 72) and other forms of still life were to be found. Reproductions of Windsor Castle have been described though research has so far failed to unearth one. One celebrated WC base is said to have carried on its interior surface the face of a politician much execrated by its owner!

Undoubtedly one of the most highly coveted designs for the WC bowl was that of the Royal Coat of Arms. Those lucky enough to receive permission to use it lost no opportunity for advertising the fact. But there were others who did not always regard the official sanction as strictly necessary. At the beginning of the twentieth century the firm of Armitage were decorating some of their bowls with the Royal Coat of Arms when they were not entitled to do so. This was discovered and a stern warning no doubt followed. One individual in the firm—the details are obviously sketchy—was determined to carry on printing the emblem. He got over the difficulty by stamping clearly above the design: 'Not by Appointment to the King' (see plate 17, p 89).

Along with these patterns matching paper holders, chain

handles and cisterns were available, making the apparatus all of one piece.

In architectural patterns geometrical designs dominated and these were often picked out in colour or in gold. Undoubtedly the most amazing of all the decorated basins ever made were the 'Lion' and the 'Dolphin' manufactured by Armitage.

In the first one, a recumbent lion with a benign expression on his face bore on his back a magnificent bowl with a bold pattern round its exterior edge (see plate 18, p 89). Hellyer mentioned this in disparaging terms because of the hygiene risks.

The Dolphin curled up into a letter S suffered the indignity of having the bowl, in the shape of a fluted shell, thrust firmly down his upturned throat (see plates 19 and 20, pp 90 and 107). There was a precedent for the use of this inoffensive creature for decoration in the WC. As noted in Chapter One, carvings of dolphins had separated the seats used by the Roman soldiers in their privy at Timgad.

Alas, the 'Lion' and the 'Dolphin' exist now only as collectors' items, and decorated ware is more often seen in photographs than in real life. All is not lost however. Armitage have made a modern attempt at decoration with their 'Decorose', where a single rose has been applied to the base of the bowl and two more decorate the cistern. Humphersons, with their recently introduced 'Tabasco' and 'Aztec' designs, perhaps come closest to the spirit of their predecessors, and their revolutionary approach has caused a minor furore in the building industry.

Quite apart from the proliferation of patterns, the Victorians were able to indulge their fancy in inventing new names for the multiplicity of products. These names must run into thousands and many firms produced gems that have sunk beneath the waves without a trace. To some extent it is possible to classify them. The more obvious have a decided 'water' connotation. Thus we find the 'Aquarius', 'Shower', 'Storm', 'Watervaal', 'Aqua Jet', 'Cascade', and 'Ripple'. Others with an aqueous connection include the ever popular river names: the 'Trent', 'New Humber', 'Ribble' and 'Vistula'.

Another class included WCs named after towns and cities. Oxford, Windsor, Malvern, Ripon, Lancaster, Blackburn and Aldwych were among those so honoured. Even this was not enough and the list was extended to include foreign names, though the bowls had been manufactured in the British Isles—Rotterdam, Paris, Hauteville, Bombay and, for the American market, Washington.

Palaces and public schools, even royal personages were put to use, including Victoria and Albert. Nor were the names of large houses entirely overlooked: Osborne, Kensington and Wellington. The 'Beaufort' paid tribute to a duke, albeit by a circuitous route, and Armitage extolled the virtues of Burlington and the empire.

Among this plethora of names were products which suggested that they possessed remarked powers. Such include the 'Rapid', 'Perfecta', 'Excelsior', 'Eureka', 'Desideratum', 'Rapidus', 'Champion', 'Inodoro', 'Directo', 'Rapide', 'Precipitas', 'Perfectus', 'Optimus' and the 'Swift'.

The British Empire was not forgotten either and India was provided with the 'Hindoostan', 'Native', 'Native Bombay', 'Sultan' and 'Oriental'. For the ordinary British market the 'Citizen' and the 'Artisan' (which curiously enough was spelt 'Artizan' for the Americans) were available.

So far the names have fitted into categories of a sort, but what can one make of the 'Clencher', 'Seestu', the 'Square Embossed Primo', 'Ryastas', 'Zeta', 'Unus', 'Arthurlie'... ?

And what is the function of a 'Frigidas', a 'Liernas', a 'United Corbel' and a 'Solidus'?

The 'Metropole', 'Closet of the Century', 'Shakespeare', 'Hebas', 'Granitas', 'Rivoline', 'Comet', 'County Council', 'Ormonde', 'Aeneas' . . . the list is never ending.

The Americans have been no less inventive in the choice of names, and research uncovers wonderful examples. Naturally enough some of them had a strong local flavour, like the 'Alaska', but the net was cast wide with the 'Siberia', 'Orion', 'Sanitas' and 'Ariel'. The most remarkable of all, rivalling the 'Lion' of Armitages, was one shaped like an elephant's head. This highly ornamented and coloured basin was designed by F. S. Moore and patented by him in 1889.

But why we ask, was there never a 'Harington'?

CHAPTER SEVEN

The History of Ball Valves and Overflow Pipes

THAT apparently simple little piece of apparatus, the ball valve, which allows a cistern to fill without overflowing, has a long history. One of the earliest references to it in print was in 1759, in the *Country Builder's Estimator*:

> Ball Cock. A water cock, the plug of which has attached to it a lever, bearing a buoyant hollow copper ball at the lower extremity. . . . When the water in the tank is drawn off, the ball sinks and opens the cock, which admits water. When this rises, the ball rises with it; in its new position closing the cock, and consequently stopping the stream of water; therefor the tank is always full, yet never runs over.

There is no reason to suppose that this was the first ball valve ever to be invented and it has been suggested that the Romans were aware of their uses over 2,000 years ago.

The early ball valves activated plug cocks which at best

Plate 17 (*above*) The wording used on an Edwardian closet made by Armitage who used the Royal Coat of Arms without official sanction

Plate 18 (*below*) The 'Lion' made by Armitage was one of the most remarkable examples of decoration in sanitary ware

Plate 19 This tastefully embossed 'Dolphin' graced many an Edwardian home

were inefficient contraptions, leaking abominably, especially under pressure. Thus cisterns meant never to run over were in a state of permanent incontinence. Once public water-supplies became common in England in the first half of the nineteenth century, some improvement became necessary. Hence the Manchester and Salford Act in 1823 which echoed many Acts dealing with private water companies:

> In order to prevent . . . waste of water . . . every person supplied with water shall provide a proper cistern to hold such quantity as shall be deemed sufficient and to provide the same with a ball and stop cock fixed to the pipe . . . preventing the water running into the cistern from running to waste when the same shall be full.

A further Act, of 1847, stipulated: 'Every person supplied with water shall . . . provide a proper cistern . . . with a Ball and Stop Cock . . . so as effectually to prevent the water from running to waste.'

Such Acts, laudable as they were, could hardly be enforced when an effective ball valve had not been invented. P. J. Davies as late as 1905 wrote: 'We now come to that part of our work where water discharges into tanks, and requires very particular attention, for without care the plumber with ball cocks will always be in trouble.' He discusses at some length the large variety of valves which at some time or another have been available for the plumber to use. Mention is made of an early invention of Bramah's in 1778, a spring valve, which he calls 'a splendid valve', but it must have left much to be desired.

The great impetus for the production of a really efficient valve came in the middle of the nineteenth century. Increasing numbers of wcs were being installed, all requiring piped water-supplies and sewage systems. It is no exaggeration to say that many private water companies found great difficulty in maintaining supplies where wastage was taking place. An

Act of 1847 introduced fines for wastage: 'Every person supplied with Water . . . who shall suffer any such Cistern, Pipe, Ball or Stop Cock to be out of repair, so that the Water supplied shall be wasted, shall forfeit . . . a sum not exceeding Five Pounds.'

The water companies on their part tried many strategems to counteract the waste. Some turned the water off all night, some went further and restricted the supply to limited hours of the day as well. Another ruse was to withdraw water altogether from specific consumers whose wastage was considered excessive. Such measures were always unpopular and came in for increasing condemnation from sanitarians and public alike. As Hellyer had pointed out in another context, it seemed scandalous to restrict the use of water in a climate as wet as the United Kingdom and this applied to a large part of North America as well.

Other factors which led to a greatly increased use of water in the United Kingdom arose out of the Chadwick Report of 1842 and the Public Health Act of 1848. At long last great importance was being paid to sanitation. But just around the corner, and serving to underline the importance of the Chadwick Report, came one of the terrible tragedies of nineteenth-century Britain. In 1849 cholera swept through the kingdom killing upwards of 55,000 people before finally being brought to a halt. An ensuing government inquiry, amongst its findings, condemned the practice of intermittent water supplies and from then on the various water companies that served cities, towns and villages were expected to provide water on a permanent twenty-four-hour service. Thus, from among a host of many other desiderata, arose the necessity of a really efficient ball valve.

The story of such a valve is a complicated one. So many people have tried to invent what at first sight seems a simple enough piece of apparatus only to run into trouble. Some were too expensive to make, some too difficult to repair and some for no apparent reason failed to gain popularity either with the water companies or the plumbers. The first real breakthrough came in 1845 when Edward Chrimes of Rotherham in Yorkshire took out patents for three new and improved valves, which for the first time gave sanitary engineers what they had so long required. Chrimes had realised that with companies now supplying water at higher pressures than in previous days the old plug cock could never be satisfactory. His valves foreshadowed the Act of 1847 which laid down that the supply of water 'shall be constantly laid on at such a Pressure as will make the Water reach the top storey of the highest Houses'.

Chrimes' valves were technically far in advance of any previously seen and water companies were not slow in recognising the fact.

The first of the three valves was activated by a float on a long arm working a cranked axle, which in turn opened or closed a valve consisting of a flat washer seating against a washer plate. Of interest, and certainly novel in float valves, was the fact that in the valve's closed position the water pressure in the pipe was directed against the back of the washer and helped to keep it firmly in place.

The second invention used an entirely different principle. A wedge-shaped arm, worked by the ball float, was responsible for lifting the valve which together with its seal was inclined at an oblique angle. The valve was moved sideways and upwards by the descent of the float, permitting the entry of water into the appliance the cistern was serving.

Chrimes' third invention used a valve and seating activated by a float working through a sharply pitched screw which kept the valve tightly closed when the cistern was full.

With these three valves in production from 1845 the plumber was able to cope with the increasingly stringent regulations of a myriad of water companies. Although later inventors sought to improve upon the original work of Chrimes, their efforts at first were of marginal importance and it was not until about 1880 that the forerunners of today's familiar valves—the Croyden, the Portsmouth and the Equilibrium types—began to be discernible. According to B. Hawkley, writing in *The Plumbing Trade Journal* in September 1961, the basic 'fundamentals of ball valve designs had all emerged' by 1877.

One of the earliest equilibrium valves, the Lambert, had emerged about 1870 and achieved great popularity because it was cheap to make and simple in operation. Ten years later new and improved equilibrium valves had come into being, including the now famous Croyden valve.

In the equilibrium valve of 1880 the rising float forces a piston into a valve seat in order to cut off the water supply to the apparatus. At the same time water flows round the opposite end of the piston which carries a washered cup. In this way pressures are equalised on both sides of the valve which can thus be used throughout a full range of water pressures.

By 1885 the popular Croyden valve was in everyday use. In this valve water from the supply pipe was admitted to the cistern through a tapering orifice which was closed by a coned valve mounted on a piston worked by the ball float. One of the advantages of the Croyden valve was its great simplicity—which meant that it was cheap to make. It earned the distinction of being the only valve chosen for inclusion in the first code of bylaws promulgated by the Joint Committee on Water Board Regulations formed in 1900. This was recognition of a high order not given to any other contemporary valves.

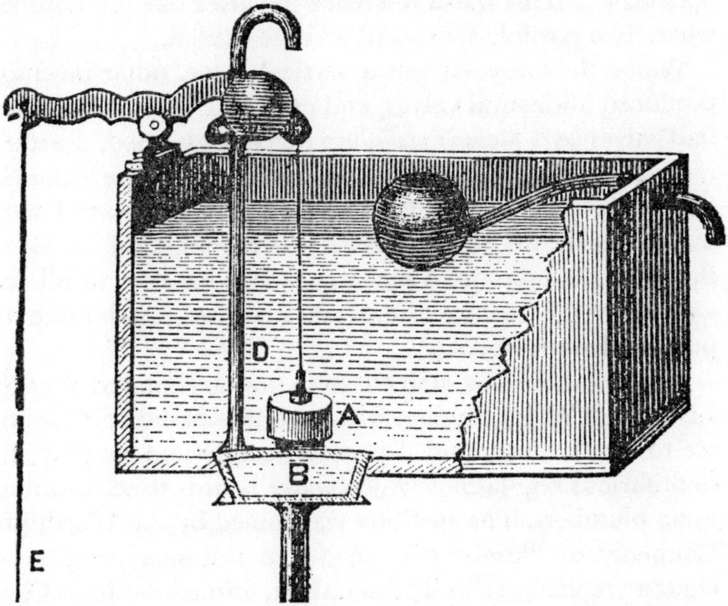

Fig 5 An early attempt to economise on water saw the invention of the Frugal Valve Cistern

By the turn of the century, however, the Portsmouth valve, which is in common use today, was challenging the popularity of the Croyden valve. There were few acceptable valves on the market and these did not undergo much basic improvement. By 1910 the Portsmouth, being cheaper, more efficient and simpler, was in the ascendancy. Standardisation of size was the next factor to receive attention. But the Croyden refused to die and was much favoured by P. J. Davies, who wrote as follows: 'A former engineer of the New River Waterworks is on record as saying: "I like . . . the Croyden, for the simple reason that with it you have only one chance of leaking; whereas, in most other of this class, there are two openings." ' This was a reference to valves like the Lambert where two possible sources of leakage existed.

Whilst the Croyden was a vertical valve, other inventors produced horizontal valves, and of these Davies said: 'These ball valves *when properly made*, are exceedingly good.' Furthermore he mentions their use on both sides of the Atlantic: 'My experience which is over forty years practical work under nearly all the different water companies in Great Britain, and some in America, teaches me that of all ball valves on the market there are none to equal those represented [in his book].'

By 1900 there was such chaos in sanitary matters that the Institute of Water Engineers had set up a Standing Committee to inquire into, and if possible, codify and simplify, the multifarious regulations which water boards tried to enforce upon plumbers. The institute was joined by the Worshipful Company of Plumbers in 1903 and this body, which included architects as well as plumbers, formed the Joint Committee on Water Regulations (JCWR). By 1908 the JCWR was in a position to promulgate its findings and published its first list of model bylaws which dealt with the use and misuse of water, the problems of contamination and similar questions. Among other things the committee laid down certain desiderata for ball valves and levers, founded on the Croyden pattern. Specific patterns of ball floats and how they should be soldered together were included, and stop taps to the cistern were regarded as obligatory.

Such recommendations, laudable as they may have seemed at the time, leave much to be desired when looked at in retrospect. Moreover, as recommendations and not official bylaws, they could not be legally enforced.

In 1911 the Local Government Board, an official body with the full backing of the law behind it, produced its own bylaws. These departed in many details from the 1908 suggestions of the JCWR and laid down such statutory requirements as the thickness of copper in the ball float, the material of the lever, the size of the orifice proportional to the pressure it was to endure, and the means whereby siphonage (and possible water contamination) was to be avoided. Bylaw 24 in the list is of interest as it paid attention to the presence and use of the overflow pipe: 'Where a ball-cock is arranged as to discharge water by a silence pipe or otherwise into any cistern below the level of the outlet to the overflow pipe, means shall be taken to prevent the possibility of siphonage of water back into the main of the undertaking.'

These regulations which were to prove so unpopular brought much condemnation from water boards all over the United Kingdom. Eventually the British Waterworks Association, in 1921, were goaded into drawing up their own model specifications through the medium of a Joint Standing Committee on Water Regulations (JSCWR). Later a joint committee representing all interested bodies was inaugurated and met in August 1923. This augmented body included plumbers, manufacturers of sanitary equipment, water engineers and architects. In little over a year they had completed their tasks, but their findings were not to survive long. By 1928 further bylaws were considered necessary and specific statutory regulations concerning ball valves were included in a host of other topics, and the Portsmouth valve at last received official recognition. These bylaws, which were produced by the Ministry of Health, laid down obligatory requirements for ball valves, as well as non-enforceable recommendations as a sort of appendix.

In 1936 the British Standards Institute turned their attention to matters of water engineering and by 1939 had managed to convince the Ministry of Health of the wisdom of using

British Standards for the various pipes, screws, fittings and so on. The war brought all this work to a premature end.

In 1945 new model bylaws were issued by the Ministry of Health and in 1946 the now famous specification for the Portsmouth valve, BS 1212, was published. This specification, largely the result of collaboration between the National Brassfoundry Association and the British Waterworks Association, made it possible, with the aid of the BSI to introduce a joint licensing scheme by which officially approved valves were stamped with the letters JCSWR and were of a guaranteed standard.

As was noted earlier the Croyden and Equilibrium valves, while not as popular as the Portsmouth, have not ceased to exist and now in its turn the Portsmouth valve is being superseded. One of the great disadvantages of valves in a hard water district is that they run the risk of sticking and jamming as a result of the calcium salts deposited in them. This led the Building Research Council to research into the possibility of producing a valve without such limitations. The result has been the Diaphragm valve. In this all moving parts are dry and are thus unaffected by water. It is also much quieter than the Portsmouth valve, a fact partly due to its having a nylon orifice. Indeed with the introduction of new plastic materials specifications for water appliances and their fittings have had to be changed. Thus in 1945 ebonite had been used for interchangeable seats in valves, only to be withdrawn two years later as it did not prove entirely satisfactory.

This is by no means the end of the story. Research continues, perfection is always being sought though never achieved. The ball valve, though apparently so simple, is still capable of being improved more than a hundred years after the patents of Chrimes.

As ball valves, however efficient, sometimes fail to perform their function, there is always the need for overflow pipes. Such pipes, as most of us know, perform the useful job of allowing water to flow away safely, should the ball valve fail to cut off the water for any reason. The law has perhaps been less specific concerning overflow pipes, but certain guidelines have been laid down. P. J. Davies mentions them in connection with the Metropolis Water Act of 1871 and recommends plumbers to fit them where they can be seen rather than allowing them to discharge into an open head; 'but if this is the case you will soon have your water man upon you'.

More recently Codes of Practice have laid down recommendations stressing the importance of having overflow pipes from cisterns arranged as warning pipes. The importance of this seems obvious enough, serving to draw attention to the malfunctioning of the ball valve so that the necessary repair can be made.

It has been customary to have overflow pipes terminating outside buildings and often discharging into nearby drains. This is not obligatory and it has been laid down (CP 310) that outlets may terminate inside a building provided they are in conspicuous places. Internal overflows may be diverted into the WC basin, over a nearby bath or wash basin, or on to the floor. This last contingency may seem strange but it has the advantage of being very conspicuous; yet one can scarcely imagine any plumber in his right mind availing himself of such an outlet except in special circumstances.

Finally some recommendations are made concerning the physical properties of the overflow pipes themselves. They should be of adequate size to prevent overtopping of the cistern if the ball valve fails. In England and Wales the pipe should have an internal diameter greater than that of the inlet pipe and in any case not less than ¾in. Scotland plays safer by insisting that the lumen of the pipe should be at least twice that of the inlet pipe and in any case not less than 1¼in. The recommended fall of pipes is in the nature of 1 in 10.

All in all today's ball valve with its attendant overflow pipe is a satisfactory piece of apparatus. It performs its purpose thousands of times a year with little thought from the user until the time comes for re-washering, and even that is a simple and quickly performed task.

CHAPTER EIGHT

Seats and Cisterns

Of the many components of the water closet none is more likely to impress itself on the user than the seat. Man has ever been alive to the attraction of making his daily visits comfortable. Certainly comfort appealed to Boswell who praised a most elegant contraption at the house of the Dutch Ambassador in Paris which had a soft padded top. Dr Johnson was emphatic in his condemnation: 'Sir, that is Dutch, Quilted seats retain a bad smell. No, sir, there is nothing so good as the plain board.'

At the other end of the comfort scale wc bowls have been produced without seats at all, designed for less fortunate folk. Thus Hellyers model, the 'Jar' (see plate 10, p 53): 'The rim is rounded and of sufficient width to allow use by any person in a sitting position without the use of a wooden seat, where desirable.' If this seemed too spartan for some institutions then models with teak or vulcanite 'haunch pieces' were available. On the other hand, there were wcs whose rims were studded with spikes so that the user was compelled to pull down the seat in order to cover them.

Neither of these models represented the ultimate in discomfort. Even today many sanitary ware manufacturers are turning out models for use in a crouching position. Such primitive appliances, which are really sunken wcs at floor level, are common enough in the East (see plate 21, p 108). Their presence is not unknown in Europe, examples being found in Switzerland, Greece, Turkey, Cyprus and elsewhere.

Wood has probably always been the favourite material for wc seats but alternatives have been tried. In 1899 Shanks introduced their 'Metallo-Vulcanite' seat in black, red or brown. This was said to have a 'core or shell of stamped steel which is covered with a substantial covering of vulcanite'. However, Shanks continued to supply seats in mahogany, oak, walnut, birch, deal and cypress.

Nine years previously, in 1890, the American plumbing industry had been similarly alive to the need for comfort and elegance in seats. The Rumsey Manufacturing Company of St Louis could supply seats in cherry, antique oak, walnut or ash with their 'Ariel' wc. Now the days of cold hygienic plastic are with us, and antique oak and cypress have disappeared from the smallest room. So too have the practical jokers, one of whom is said to have made a wc so that when anyone sat upon its seat, a hidden device played 'God save the Queen'. For a patriotic Victorian Englishman the use of such a machine must have caused a conflict of honour and necessity—it was unthinkable to sit while the National Anthem was played! The approach to the subject of seats has certainly been varied, and where comfort is concerned on a cold day, the well-known aphorism is not to be despised: a swinging chain means a warm seat.

But if a seat is important perhaps it is sensible to consider what supports it. An unfortunate experience was recorded in the *Daily Telegraph* of 12 March 1970:

> When Mr Gordon McLeod sat on the lavatory at Glasgow public baths the front of the pan broke away and he fell on to jagged china, suffering serious cuts. He received 51 stitches in his buttock wounds, spent three days in hospital, and was off work for two months.

In the development of the wc the importance of both seats and cisterns was early recognised. The need for a cistern that could fill automatically and then cut off the water supply was dependent on the invention of a ball cock or similar device.

For some reason the subject of cisterns, or water waste preventers was one of great passion to S. S. Hellyer. It was a topic to which he was to return time after time, his eloquence, rhetoric and vehemence rising and falling like a copper float. In his book *The Plumber and Sanitary Houses* he spends almost a whole chapter on the subject of cisterns:

> It ought to be laid down as an axiom that no supply of water is adequate to the purpose of keeping a wc, or any such fixture with its belongings, wholesome, unless it exceeds in force and volume the body of the polluted matter discharged into such fittings.
>
> The quantity of water allowed by the various water companies and made legal by Act of Parliament (Metropolis Water Act 1871) was two gallons for each use of the closet. Two gallons to carry away a deposit in a wc through scores of feet of soil pipe and drainage, flush these pipes out after its passage and cleanse the whole! This is about as difficult as making bricks without straw.
>
> Let us look at things practically i.e. with an open eye. Suppose it takes two gallons for a wc in the basement (where we guess its discovery to have been made) to cleanse the closet after usage, wash out the trap, and deliver the deposit into the sewer, how much will it take for a closet fixed on the second or third floor at the rear of the building with about fifty feet or more soil pipe and fifty to sixty feet of additional drainage? Why it would require this quantity to wet the pipe and two to three times the quantity to keep it all clean and wholesome.

> Every sanitarian should lift up his voice against the limitation of water to such sanitary fixtures and never cease crying, like Oliver Twist, for more water until a quantity double or even treble the present amount is allowed for wcs.

Even all this is not enough. Hellyer goes on to describe experiments with a glass panel let into the side of a wc trap and of the abominable results of a mere 2 gallons. Worse is to come. He condemns people who are too careless or too idle to give the wc a proper flush, and suggests that a neat white porcelain tablet lettered in gold should be placed, where a person using a valve closet could see it:

AFTER USING THE WC PULL UP THE HANDLE TWICE, AS FAR AS IT WILL GO AND HOLD IT THERE A SECOND OR TWO EACH TIME.

Finally there is a bitter condemnation of the water companies, which Hellyer regarded as monsters to be attacked at every opportunity.

> It is monstrous that water companies should have the power to enter a man's house and interfere with his comfort by restricting the supply of water to his wc. For a small sum of money per annum the London Water Companies allow a man to have a bath in his house where he may consume 50 gallons a day but they will not sanction more than two gallons for the use of the wc.

P. J. Davies had also a great deal to say on the subject of flushing cisterns though the emphasis of his remarks was somewhat different from Hellyer's:

> Plumbers cannot give too much attention to the thorough flushing of closets, for upon this point, . . . depends the health of the poorer class home.
>
> I have for many years made this branch of our trade my particular study. . . . I shall lay before you all those worthy of your attention, such as Jenning's, Tylor's, Lambert's, Boulding's, Guest & Chrimes etc., so that you may judge. . . .

In the earliest days of the wc the main function of a cistern was to contain water to enable flushing to take place without consideration for the amount used. As society became more sophisticated and water was being used in ever increasing quantities, it became necessary to restrict this usage and the invention of the water-waste preventer (WWP) was a natural corollary. Today the word cistern has come to mean a WWP and the term which so angered Hellyer and others like him has passed out of common parlance. For this reason the term cistern is used generally in the text in its modern meaning.

It must not be supposed that Hellyer and Davies were alone in their regard for the importance of the cistern. Shanks of Barrhead, Glasgow, ever keen to keep abreast of events, had worked hard and successfully on the development of their Leverne Cistern which could be fitted with Dicksee's Patent Automatic Flushing apparatus. By 1895 Shanks were claiming a silent low-level cistern on which 'an eminent sanitary engineer has made experiments, and has calculated that the discharge . . . rushes through the soil pipe in three times the volume of that of an overhead cistern'.

In 1899 the catch phrase of the sanitary engineer seemed to be: 'A good cistern makes a good closet.' As though seeking confirmation of this homely aphorism Shanks announced another and more splendid device, the 'Aqua-Jet cistern', more costly than the 'pull-and-let-go' chain type at the quite high price for those days of £9. This cistern had a press button action resembling an electric bell in appearance and found favour with certain Indian princes. The variations were many: cast-iron, vitro porcelain, embossed, ornamental, nicely japanned, copper-lined mahogany and walnut. Shanks' fascination with cisterns continues: 'When the cistern valve is lifted, the contents of the cistern are precipitated with one large gulp into the basin.'

There were single-valve cisterns, double-valve siphons,

plungers and diaphragms, frugal valves, waste-not valves, and a vast gallimaufry of different sizes and models to suit a variety of water boards (see fig 5 and plate 26, pp 95 and 110). Despite this Shanks boldly stated, 'Truly a need of the times a silent, Fullway, High-Pressure Ball-Valve,' and with that we can leave Barrhead to its own devices and consider what was happening elsewhere around the turn of the century.

The Albion Clay Company in the Midlands boasted a cistern patented by Winns which could be 'supplied with improved internal fittings of brass and ironwork coated with Dr Angus Smith's solution, passed by Southend Water Company'. Who, we wonder, at this time in history was the inventive doctor who succeeded in perfecting a solution which satisfied the requirements of Southend?

By 1896 T. C. Brown-Westhead, Moore & Co Ltd of Cauldron Place, Hanley, who were early pioneers in the export of sanitary ware to America, were turning out thousands of porcelain cisterns. Their catalogue for that year shows 'Cauldron Ware' cisterns decorated with flowers and more formalised designs; such cisterns were in common use well into the twentieth century. A fine example of a decorated cistern, with a picture of wading birds on it, exists today in the Shire Hall in Monmouth and there must be many similar.

Another man whose name crops up in the development of WWP devices in the last century is that of Thomas Crapper, a London plumber, to whom two of the Humpherson sons had been apprenticed. Crapper patented a valveless WWP which claimed to be 'a certain flush with an easy pull'. He was well thought of in London in his time and his work was good enough to draw the attention of the royal family. When Sandringham became a royal residence Crapper was asked to undertake the sanitary installations and remnants of his work remain to this day.

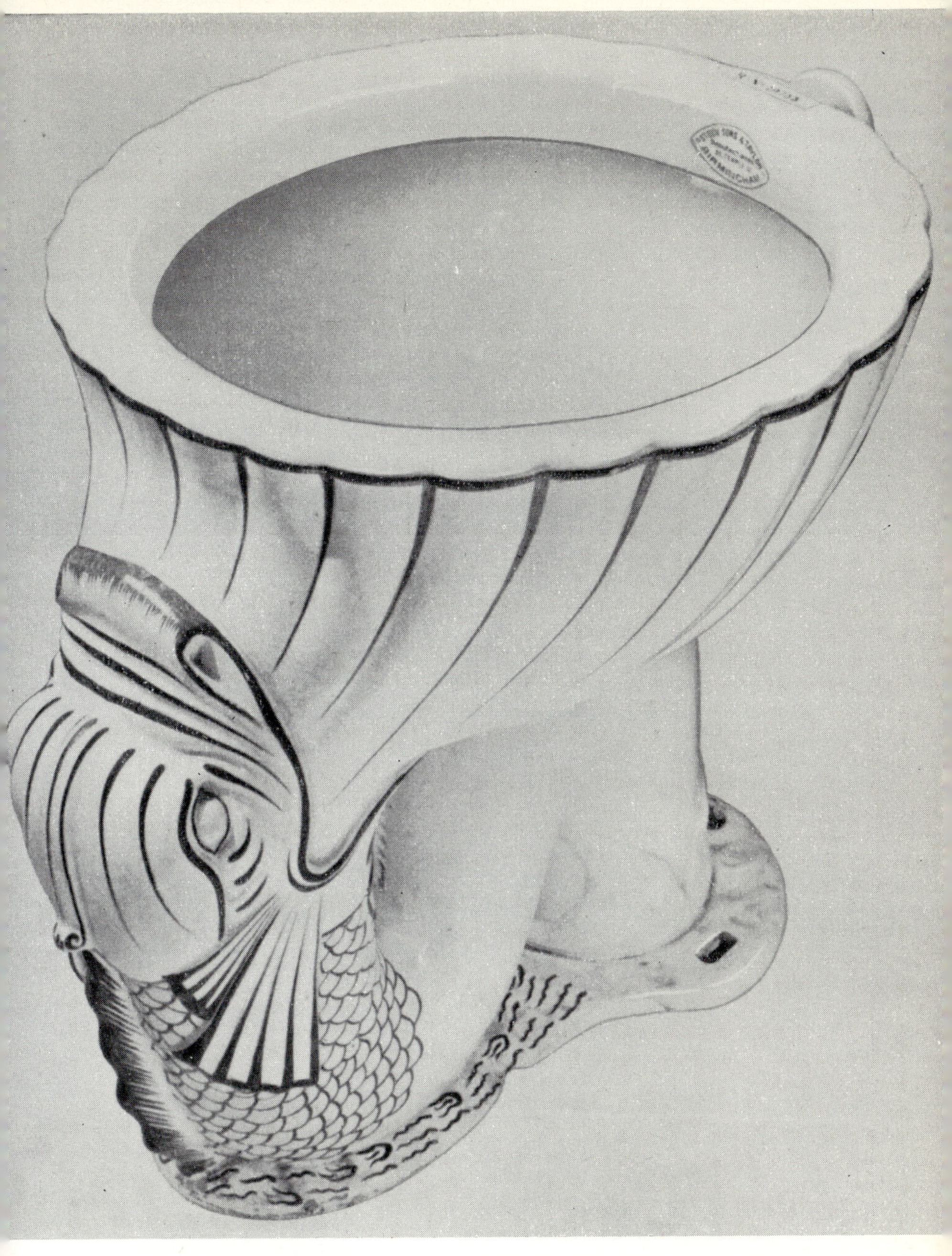

Plate 20 An example of a fluted, embossed and coloured basin, this 'Dolphin' was decorated in blue

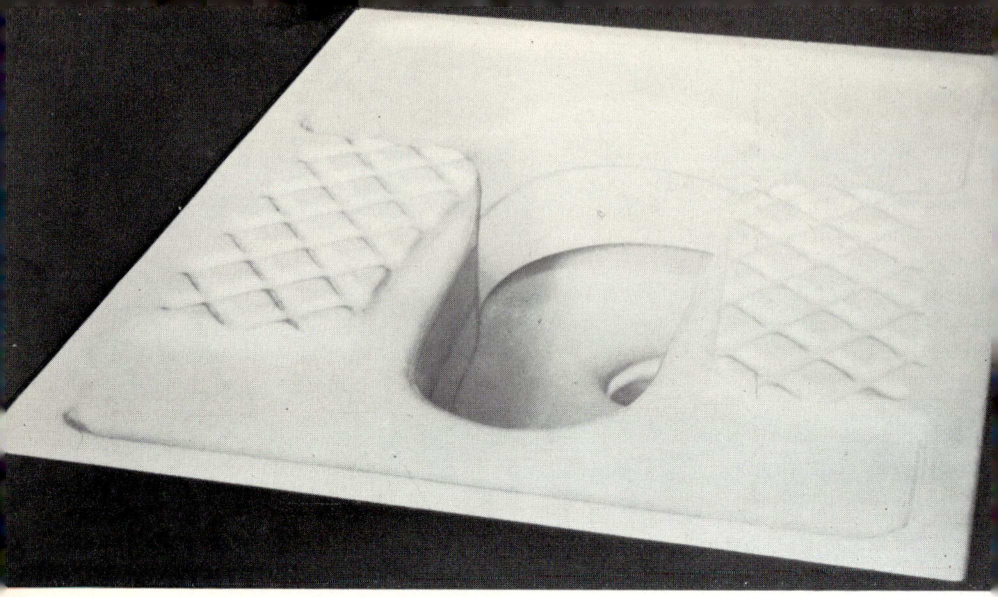

Plate 21 (*above*) In many countries, especially in the East, WCs are not fitted with seats. The Armitage 'Artesia' is a modern design

Plate 22 (*below*) The 'Unisyla' of Armitage Shanks shows a modern double-trap siphonic suite

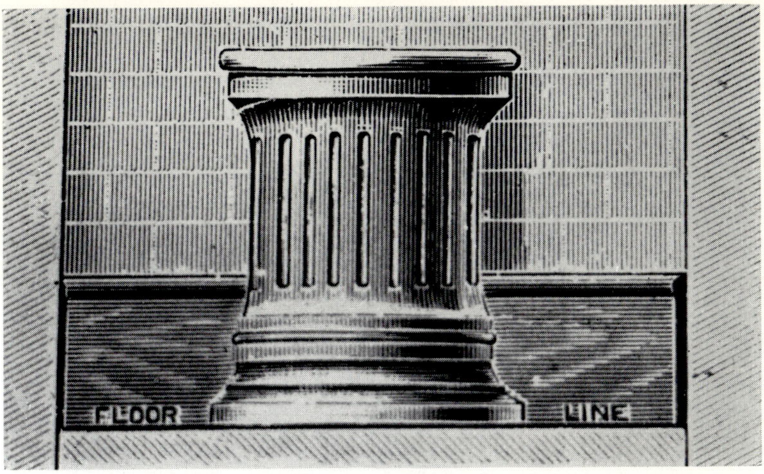

Plate 23 (*above*) This simple and functional pedestal was intended for use with Ducketts slopwater closet

Plate 24 (*below*) The famous 'Beaufort' of Humphersons claimed to be the 'Original Pedestal Wash-Down Closet'

Plate 25 (*above*) Jennings' siphonic 'Closet of the Century' created a sensation when first introduced

Plate 26 (*below*) Stones were early manufacturers of cisterns and their single-action model was deservedly popular

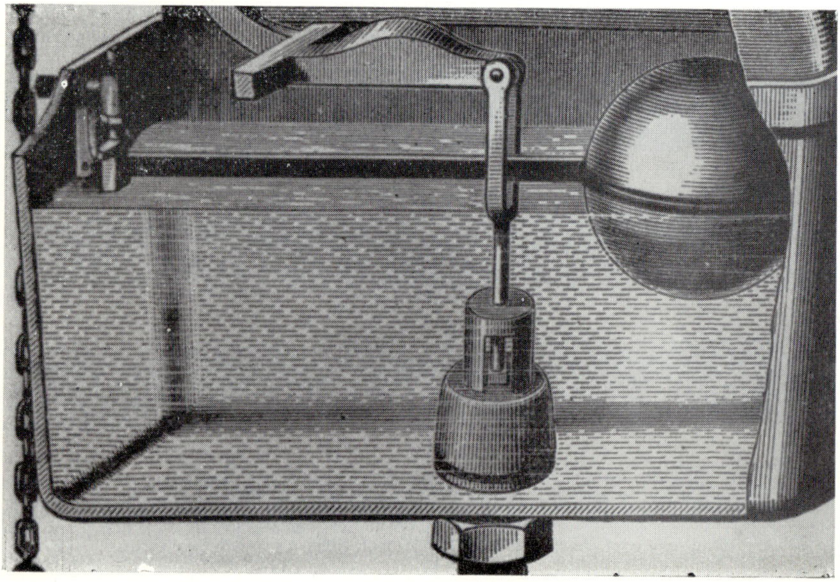

In the early history of the cistern a reasonable head of water was considered necessary for adequate flushing of the wc to take place. As improvements in design came along it was found possible to bring the cistern and bowl nearer together. The result was the low-level or close-coupled suite which is familiar to most of us today. A typical example from 1930, the Armitage 'Universia' is shown on p 128. A more modern version, manufactured by Armitage, is shown on p 108.

It might have been thought that a device that used the waste water from the kitchen sink to flush the wc would have found favour with the water boards. Such 'automatic slop water closets' were turned out in their thousands by Ducketts of Burnley and many are still in use (see plate 23, p 109). They were particularly useful where the wc was sited some distance from the house and they found their greatest application in the industrial North of England, where the little house at the bottom of the yard was commonplace.

The Automatic Slop Water Closet included a pivoted self-acting tipper which could be filled with waste water from the kitchen sink. When the tipper was full it overbalanced, sending its contents rushing down a pipe to the wc some distance away and flushing it clear. Ducketts in 1913 were advertising several models in a price range of £1 10s to £2, with many optional extras. An ornamental pedestal, amber outside and white inside, was a mere 8s, and a varnished hinged pine seat 2s 6d. Tippers were of 2, 3 or 4 gallon capacity and could be fitted either close to the outlet of water from the sink or close to the wc pedestal.

Such a clever idea was not confined to the single closet. Multiple units were available for such places as factories, weaving sheds, asylums and the like. A unit of five closets cost £6 5s, whilst one of nine closets cost £10. In these closets two tippers were advisable, one at each end of the range and of increased capacity. Two 5 gallon tippers were advised for flushing up to nine closets. Beyond that 10, 15, 60 and even 100 gallon tippers were manufactured. S. S. Hellyer would certainly not have quarrelled with such lavish quantities of water. But not everybody was of the same opinion as Hellyer and it was precisely this use of water which led in 1892 to a legal argument between the corporations of Liverpool and Chorley, as the *Liverpool Courier* of 28 May 1892 reported:

> In the Queen's Bench Division yesterday, the case of Evans v Gornall again came before Mr Justice Mathew and Mr Justice Wright, in the form of an appeal . . . from the decision of certain justices of Chorley, raising an important question as to whether the inhabitants were entitled to use their waste water for the wcs or any other purpose.
>
> Mr Moulton QC with whom was Mr Forbes Lancaster, continued his argument in favour of the appeal. He said that although the sum of money involved in this particular case was very small, the point raised was of considerable importance. The contention of Mr Gornall was that having once used water for domestic purposes he was entitled to use it in any way he liked before allowing it to run into the sewer. He (the learned counsel) submitted that the water was supplied expressly for domestic use and there was no right on the part of the customer to use it for any other purpose. If the use of the apparatus employed by Mr Gornall became general, not only would the special charge for water for wcs be avoided, but the water which was supplied without limit for domestic purposes would be so lavishly used that the difficulty of the supply would be enhanced. Apart from mere waste, they objected to the action of Mr Gornall on sanitary grounds. It was true they were not a sanitary authority, but they objected to any connection between such apparatus and their pipes.
>
> Their lordships, without hearing counsel for the respondent, dismissed the appeal with costs. Mr Justice Mathew said, in his opinion, what was contemplated by the section was a misuse of water. Mr Justice Wright was of the same opinion. The sole object of the section was to prevent a person taking more water

> under the colour of domestic use than was reasonably wanted for that purpose.

Only three years later the following facts concerning Burnley were published and the popularity and effectiveness of Ducketts products vindicated:

> Within a few years it is expected that every town will have water-borne disposal. In Burnley there are 100,000 inhabitants in 20,000 houses. In 1888 there were 6,586 wcs and 9,009 pails, cesspools etc. In 1895 there are 16,842 wcs (of which 9,071 are slop water closets) and 3,816 pails, cesspools etc.

Quite apart from the hygienic benefits of waterborne sanitation, Ducketts were able to make an unanswerable appeal to the hard-headed town councillors. The cost per annum of a slop water closet was reckoned at 1½d, and at 5s for pails. Ducketts also emphasised the point that slop-water closets saved water.

The efficacy of flushing cisterns and their ability to dispose of a large variety of unusual materials may be judged by some of the tests to which they have been put. A classic example, often quoted, concerns a remarkable performance by a Jenning's 'Pedestal Vase' which won a gold medal at the Health Exhibition in London in 1884. This voracious machine consumed, with a mere 2 gallon flush, ten apples, one flat sponge, a smearing of plumbers' smudge (whatever that may be) and, for good measure, four pieces of paper.

Rival tests by other firms were inevitable. In 1890 Ducketts were announcing, 'the discharge from the tipper enters the pan at a tangent and after circling round it is directed into the centre by a stop; this gives a powerful whirlpool action . . . in an experiment a large quantity of dry grass, enough to fill the basin, was completely removed by one discharge of the tipper'.

How fitting such a test may have been perhaps escaped the notice of the instigators. In the Middle Ages, before the common use of paper, the completion of the toilet was often by means of mempiria, balls of hay rolled up for this special purpose. By 1890, however, cut-up sections of newspapers and similar printed material were in common use and mempiria, like gongfermors, were obsolete bygones of a departed age.

Apples and hay are still with us but man in all his cunning has not designed a more fiendish article for disposal by water than the modern ping-pong ball. Some cisterns are able to cope with such a challenge. Surely this is the ultimate!

That the cistern is regarded with much seriousmess is shown, at least in Britain, by the existence of a Flushing Cistern Makers Association. Steel, plastic and rubber have been added to the traditional materials of former days and research continues with great vigour and intensity. The differing demands of hundreds of water authorities create many problems which cistern manufacturers strive hard to overcome.

The 'Burlington', the 'Jubilee', the 'Birmingham Test Well Bottom', the 'Si-Flush', the 'Niagara', the 'Overhead Lever', the 'Internal Lever', the 'Rubberline', the 'Dolphin', the 'Japkap' . . . they flow unceasingly.

Perhaps Indian princes are at a premium and the democratic needs of the world are more readily appreciated. The 'Mosquito Proof Cistern', which has been approved by the Bombay Municipality, 'is designed to exclude all insects using water as a breeding ground and to prevent contamination'.

It has been a long history from the 'Native Bombay' wc of Johnson Bros (Hanley) Ltd to the 'Mosquito Proof Cistern' of Thomas Dudley Ltd. The quest for ever improving standards of sanitation has yielded a rich bounty in human health and happiness by preventing much disease; yet its pioneers are largely unknown and unsung.

CHAPTER NINE

The Water Closet in Country Houses and Hotels

IT is difficult to know at what stage in history the WC was regarded as essential in country houses. According to Hellyer WCs were in general use by 1814 but he gives no data to support this statement. There is, however, clear evidence that they were becoming less of a luxury as far back as 1775, when Cummings patented the first one. Even he pays tribute to earlier ones, referring in the text of his patent to a 'water closet upon a new construction', thereby implying that WCs as such were already in existence.

Before this date there are few references in contemporary writing to WCs for the simple reason that few existed. When some enlightened builder or architect fitted one in a house the fact was soon advertised because of its unique character. Thus Robert Clive, the hero of India, when he had Claremont House in Surrey rebuilt by Capability Brown between

1770 and 1774, had water closets installed. This was so rare an event for the times as to invite comment. It is also interesting to note that these closets pre-dated the Cummings' patent, thereby lending support to his remarks.

Hellyer in his book *Plumbing*, published in 1893, gives additional evidence by describing a WC actually in his workshop at the time of writing and of which he says: 'As far as I can make out this represents the best form of water closet used in England about a century or a century and a half ago.' This he describes as a simple plug closet made of marble with a plunger plug in the base, a service pipe to bring water into the bowl, and an overflow pipe high up at the back. He further discloses that he discovered two similar ones at Osterly House, twelve years previously (c 1881). Of these he says: 'A niche in a fair sized room was formed to receive the marble closet-pan, and a door, shutting up close to the seat, hid the whole arrangement from sight. A lead soil-pipe was connected with the outlet plug waste of the pan, and continued from it to the drain, which was brought into the house to receive it.'

Other early houses which departed from the outside privy were those of Keddlestone and Holkham. At the latter the owner, William Kent, had a two-seater water closet fitted in the hall of the mansion though this has been described disparagingly as being 'in a tiny windowless room'. At Luton Hoo, Robert Adam was much more generous and had no less than four WCs fitted. These were supplied with water, admittedly at a very low pressure, as befitted the plug cock valves and pipes of those days.

Despite the presence of such appliances in country houses Hellyer is not too enthusiastic about them. In the fifth edition of *The Plumber and Sanitary Houses*, published in 1893, he has this to say:

> In Country Mansions, it is not at all an infrequent thing to find a water-closet, sink or lavatory so situated that a little fortune has to be spent on it to make it and its belongings sanitarily perfect. In such case the 'fitting' should be done away with, or a better place found for it. It is better that a person should be his own carrier, and walk a few feet more to reach a water-closet, than put a strain upon the drain which it would not bear without becoming offensive.

Not all big houses offended and we are told that 'the plumbing work at Marlborough House, Buckingham Palace and in scores of noblemen's and gentlemen's houses in various parts of the country, has been reconstructed on the principles laid down in the first edition of this book'.

Certainly when Buckingham Palace had first been built for George IV by Nash its sanitary arrangements were very poor. In many ways Nash's efforts were a disaster and though it cost about £700,000 by the time Queen Victoria came to live in the palace she found it a very trying experience. Practically none of the rooms containing water closets was ventilated, with results that can be imagined; drains were faulty and even her private apartments were ventilated through the common sewer. By 1842, the year of the *Chadwick Report*, Prince Albert had persuaded the Queen to have modern water closets fitted in Windsor Castle and it is quite possible they were 'Optimus Valves'. Thus Hellyer's reference in 1893 suggests that the same transformation had taken place in Buckingham Palace. The firm of Hellyers was well known in royal circles and their services were called upon: 'In 1897 the temporary arrangements sanitary were placed in our hands on the occasion of Queen Victoria's Diamond Jubilee; and again on the occasion of the coronation of King Edward VII we were so honoured.'

Reference is also made by P. J. Davies to the fitting of a siphonic closet in Kensington Palace. This WC invented by

Davies was named the 'Kensington' for this reason and claimed to be almost silent in action.

Slowly modern sanitation was coming to all classes of housing, filtering down from the highest in the land. In *Early Victorian Architecture in Britain* by H. R. Hitchcock, mention is made of a water closet in a house built in 1835. The famous architect A. N. Pugin, on the occasion of his second marriage, built St Marie's Grange near Salisbury for his new bride. This was a most curious house with a chapel on the ground floor which was reached by a drawbridge, but perhaps even more remarkable was an upper floor containing three bedrooms reached by a spiral staircase. The bedrooms all opened one into the other and a water closet, most probably a valve closet, opened off the middle one. This WC was placed in a small rectangular turret which had a separate roof over it and housed additional WCs on other floors opening into the kitchen and the library.

Just as the *Chadwick Report* had stirred Prince Albert into action, so a spurt was given to sanitation generally throughout the United Kingdom. In 1845 the Society for Improving the Conditions of the Labouring Classes took upon themselves the task of erecting dwellings of different sorts which would serve as models of their kind. Thus in Clerkenwell terrace houses were built with the unusual convenience for those days of indoor sanitation. These houses and others were stated to be 'models of the different kinds of dwellings they would recommend for the industrious classes in the populous towns'. Later the society turned to building lodgings for single men containing dormitories for 104 occupants and with facilities for washing and water closets.

By 1856 Hunt and Stephenson had built flats in Vauxhall Row, Lambeth, in each of which was fitted an indoor WC; by this time it is likely that such an appliance was regarded as a necessity in new houses in the city.

About the same period MacKenzie had built on the south side of Victoria Street rows of houses which contained separate WCs for the servants and for the family. Such designs led, of course, to manufacturers of WCs producing cheaper models especially for the servant classes.

But if indoor sanitation was regarded as necessary in urban areas it was more usual to find the WC sited in the back yard of the house in the country. Thus in 1857 when Roberts built a block of houses for the Windsor Royal Society in Windsor all the WCs were outside the houses.

Inevitably, as water closets proliferated certain desirable standards began to emerge. G. A. T. Middleton in his book *The Drainage of Town and Country Houses* (1903) had much to say on the subject:

> 8. With respect to water-closets, earth-closets, privies, ashpits, and cesspools in connection with buildings.
>
> Every person who shall construct a water-closet or earth-closet in a building shall construct such water-closet or earth-closet in such a position that one of its sides at least shall be an external wall.
>
> 9. Every person who shall construct a water-closet . . . whether the situation . . . be or be not within such building, shall construct in one of the walls . . . a window of not less dimensions than 2ft by 1ft exclusive of the frame, and opening directly into the external air.
>
> He shall, in addition to such window, cause such water-closet . . . . to be provided with adequate means of constant ventilation by at least one air brick built in an external wall of such water-closet . . . , or by an air shaft, or by some other effectual method or appliance.
>
> 10. Every person who shall construct a water-closet in connection with a building, shall furnish such water-closet with a separate cistern, service-box, or flushing-box of adequate capacity.
>
> He shall furnish such water-closet with a suitable apparatus

for the effectual application of water to any pan, basin, or other receptacle with which such apparatus may be connected and used, and for the effectual flushing and cleansing of such pan, basin, or other receptacle, and for the prompt and effectual removal therefrom of any solid or liquid filth which may be from time to time deposited therein.

He shall furnish such water-closet with a pan, basin, or other suitable receptacle of non-absorbent material, and of such shape, of such capacity, and of such mode of construction as to receive and contain a sufficient quantity of water, and to allow all filth which may be from time to time deposited in such pan, basin, or receptacle to fall free of the sides thereof, and directly into the water received and contained in such pan, basin, or receptacle.

Evidence of the widespread use of waterborne sanitation is contained in the impressive list of houses, clubs and institutions to be found in Dent & Hellyers Catalogue for 1907. Brief reference has been made elsewhere to the buildings in which the 'Optimus Valve Closet' had been fitted and the list is worth reconsidering. It includes the Houses of Parliament, the Royal Court of Justice, the royal palaces of Windsor, Balmoral and Buckingham, the Foreign Office, the Admiralty, the Junior Army and Navy Club, the National Liberal Club, and many others. Among the illustrious names are found the King, the Princess Henry of Battenburg, the dukes of Norfolk, Richmond, Gordon, Marlborough, Portland and Bedford who were all daily worshippers at the 'Optimus' shrine. And if Dent & Hellyers were supplying the aristocracy, doubtless other manufacturers of WCs were too. In fact there is evidence of Thomas Crapper being asked to undertake the sanitary installations at Sandringham when it passed into royal ownership.

The fitting of WCs in hotels followed roughly the same pattern as in country houses. In the preface to the first edition of *The Plumber and Sanitary Houses*, Hellyer writes:

> Under the old system these sanitary conveniences generally *advertised themselves*, especially in hotels and places of that kind; and all that one had to do in such buildings was to follow the scent, like hound after a fox, by the dictates of an organ which is very useful, but which one does not care to abuse in such a way, for, to say the least, it is an offensive way of following up a thing.

By 1907 the list of palaces, clubs and large houses which boasted the presence of an 'Optimus' had been augmented by the inclusion of several well-known hotels, such as the Langham, the Buckingham Palace Hotel, the Savoy and the Berkeley. Certainly it would be safe to assume that by now no hotel in the United Kingdom is without at least one water closet on its premises. Yet the primitive may linger on a long time. The writer remembers visiting a small country pub in Oxfordshire sometime during the early 1940s and asking the whereabouts of the urinal. 'There ban't one', he was told, 'us allus uses the garden.' Johnson's days of plucking a rose were not quite at an end. There was presumably a WC in the house for the use of the residents but the customers were not so well provided.

It is probable that the development of the water closet, such as it is known today, was in large measure the result of British endeavour, though as we have seen, the Americans were not slow in following the lead. Elsewhere the benefits of sanitation were slow in coming. Thus in Paris as early as 1513 it had been laid down that every house should have a privy; yet by 1700 many were still without. The police were empowered to lock up the houses of any citizen who failed to provide one.

On the other hand privies seem to have been commoner in the larger German towns and cities than they were in Paris, though their condition left much to be desired. Beckman, writing in his *Inventions*, says of Berlin in 1846:

> In most of the houses small closets are located on the landings of the stairs, which require to be emptied every other night, to the no great satisfaction of the olfactory nerves. Nor are the streets kept in a very proper state, large puddles of filth being allowed to collect before the doors even of the best houses, and which especially in the summer months, diffuse a most horrible stench.

But if this was not enough, conditions in Spain were even worse. In *Cook's First Voyage* we are told, 'the residence of the king of Spain was destitute of this improvement, at the very time that the English circumnavigators found privies constructed, in the European manner, near the habitations of the cannibals of New Zealand'.

On casual consideration it would not appear likely that Yugoslavia would rank high on the list of countries possessing early examples of water closets. Yet according to Rebecca West such pieces of apparatus were known there as far back as 1790. Speaking of a visit to a ruined Pasha's palace in Bardovtsi she writes, in *Grey Lamb and Black Falcon,* that there were bathrooms and water closets of a standard unlikely to be found in the houses of England or America at that time.

The history of the water closet in America is a long and illustrious one, but the name of the first American to have a WC remains a mystery which perhaps will never be solved. However, there is no doubt that when the well known and much revered poet Henry Wadsworth Longfellow had one installed in his New England house in 1840 the water closet was well on the way to acceptance in America.

This single act did more to popularise the water closet than perhaps anything else. Just as what Queen Victoria and Prince Albert did in the United Kingdom quickly became the accepted code of behaviour, so the Americans followed their famous citizens, be they poets, politicians or presidents.

Apart from country houses and hotels there is evidence

that the large department stores were early to appreciate the benefits of sanitation for their staff and customers alike. Harrod's of Knightsbridge, London, perhaps one of the best-known stores in the world, were early pioneers in this field. Built around the year 1912 the store rapidly set new standards in luxury both in its fittings and its products. The Ladies' Retiring Room of those days illustrates an opulence which has long since disappeared (see plate 27, p 127). Elaborate marble wash bowls, ornate mirrors, panelled marble walls, tiled floors and an ornate ceiling remind us of a luxury to which few can aspire today. There were six cubicles with solid wooden doors intricately carved and it is probable that the WC bowls inside were also the most expensive of their kind. Though the room itself has survived into the seventies, the original WC bowls have been replaced by the most utilitarian of white models, no doubt functional and efficient but by no means beautiful. Nor has research succeeded in discovering what the originals were like.

CHAPTER TEN

The Water Closet Today and in the Future

IT would be idle to suppose that the ultimate in indoor sanitation has arrived. The old pan closet has probably ceased to exist both in America and the United Kingdom. The valve closet has also suffered a decline, albeit a lesser one, though some remain in use. Valve closets are also found in railway trains.

With wash-down, wash-out and siphonic closets the picture is very different. The wash-down quickly established its supremacy and its popularity has never waned. Although all three types of closet are still being manufactured, a new invention threatens the supremacy of the old faithfuls. Just as Hellyer was unable to paddle against the stream of the water boards, and was forced to fit water-waste preventing cisterns, so today's water engineers are having to take note of the ever increasing demand for water all over the world. The abbreviation WC is being replaced by VWC, or vacuum water closet, which is the invention of a Swedish engineer, Joel Lilljendahl, who has produced a machine which can function on a mere litre of water. This is the equivalent of 1¾ pints. An annual saving of between 23,000 and 40,000 litres (5,000–8,500 gallons) per person is claimed. With an average of 7,500 gallons per person per annum London would save 75,000 million gallons and New York 90,000 million, and some of the problems of water shortage would disappear.

Quite apart from this gigantic saving of water further economies can be made with the Lilljendahl system. It is quite easy to keep the so-called 'black' water from the vacuum WC separate from the 'grey' water of domestic use—that from the bath, shower, wash basin and other such appliances. Such grey water can be easily and cheaply purified for further use, or it can be used untreated for irrigation purposes.

The vacuum WC is activated by a small pump which keeps the evacuation pipe from the WC bowl at a pressure of about half an atmosphere. Flushing, which is started by a press-button control at the back of the bowl, admits air at normal atmospheric pressure into the system and the detritus in the form of a 'liquid-plug' is thus propelled efficiently and speedily to a collecting tank. From here it can be transferred to a chemical sewage plant for treatment with a lime-reactor and an ammonia expeller. Such plants can be kept small because of the small quantities of water involved.

The VWC has been in use since 1959 and countries with a water shortage have not been slow in recognising its advantages. The first large-scale installation in regular use, as one might expect, was in Sweden but further examples are to be found in the Bahamas and in Mexico City and there is no doubt that the popularity of such a system will increase.

Plate 27 The opulent Edwardian Ladies' Rest Room at Harrods had water closets with magnificently carved rosewood doors

Plate 28 (*left*) An early example of a close-coupled WC suite, the 'Universia' dates back to about 1930

Plate 29 (*left*) This wall-hung toilet suite, the 'Belgrave' by Armitage, was designed for the Cunarder *QE 2*

Apart from the development of the vwc, indoor sanitation in all its wider aspects has become much more sophisticated. The outside water closet, perhaps of the slop-water variety once regarded as the acme of perfection, does not meet to-day's requirements. Nor is the weekly bath in front of the fire, which was all that many Victorians knew, enough. Bathrooms have become extensions of the home, heated and carpeted to a state of luxuriousness that would have seemed sybaritic a generation ago. Along with these changes in Great Britain has come the prospect of entry to the Common Market. Many manufacturers of sanitary ware have abandoned the traditional measures of feet and inches and gone over to metric modules. Confining the size of the various fixtures of the bathroom, as well as tiles and windows, to multiples of metres, has led to much simplification in this field. The growing popularity of corbelled wcs which are entirely free from the floor is another discernible trend.

Armitage-Shank with their 'System 7' have pursued the metrication idea following the announcement of the Government's Metrication Programme. Introduced to the plumbing trade in 1971 the system has since become deservedly popular.

Rationalisation of panel sizes has led to the manufacture of pre-plumbed units, the fixing of which is easy compared with the difficulties which must have been met by those stalwart pioneers P. J. Davies, S. S. Hellyer and their contemporaries.

A typical example of the modern corbelled unit is the 'Walden' of Twyfords. It has a neatness which was entirely lacking in earlier models. The Armitage-Shanks suite 'Belgrave', originally designed for the liner *QE 2*, provides a further example of this type (see plate 29, p 128) which is familiar to many transatlantic travellers.

Concurrently with these changes has come the search for still better materials for WC basins. Ideal-Standard claim to have developed a new ceramic material which much resembles marble in appearance. Having the feel of real stone, it is a compound of clays and colours carefully mixed before being fired. One of its main disadvantages seems to be the high cost of production but the luxury article will always find a buyer.

Another departure in the history of the water closet has been the flight from white and the return to colour. Ornate patterns such as the Victorians loved are no longer fashionable but the single overall colour, in a variety of shades, is commonplace. Some of these have names almost as exotic as the flower motifs they have replaced. Among them are Coral, Primrose, Sky Blue, Turquoise, Avocado, Sun King, Savannah, Orchid, Midnight, Pampas and Shell Pink.

Two further recent inventions deserve mention. The Swedes have perfected a method by which the use of toilet paper will no longer be necessary. The performer on this amazing machine has his rear quarters washed by a jet of warm water after which he is dried by warm air. This so-called 'Clos O Mat' has obvious advantages for disabled people, in particular, for whom toilet facilities constitute a special problem.

The other worthwhile invention is that of the 'Dual Flushing Cistern' which is capable of delivering either one or two gallons of water as required. Like the vacuum WC it is useful in areas of water shortage.

These inventions look to the future. For most of us the wash-down and siphonic closets will be our daily companions for many years. They have had a long run and their further use seems assured. Inevitably they must disappear, as Sir John Harington's 'Ajax' did, but their contribution to civilised living has been great and they deserve honourable mention in the list of inventions of the Western world.

Meanwhile the good work started by Sir John Harington nearly four hundred years ago will no doubt proceed.

Bibliography

Batchelor, E. A. 'Sanitary ware for Europe?', *Ceramics,* 18 (1968)

Beckman. *Inventions* (1846)

Beresford-Evans, J. 'Towards better ceramic sanitary ware', *Design* (1954)

Burton, Elizabeth. *The Georgians at Home* (1962)

Church, Roy E. *Economic and Social Change in a Midland Town. Victorian Nottingham* (1966)

Davies, P. J. *Standard Practical Plumbing* (three vols) (1905)

Dyos, H. J. *Victorian Suburb—a Study of the Growth of Camberwell* (Leicester, 1970)

Eassie, W. *Sanitary Arrangements for Dwellings* (1874)

Finer, S. E. *The Life and Times of Sir Edwin Chadwick* (1952)

Fosbrooke. *Encyclopedia of Antiquities*

Frankfurt and Pendlebury. *City of Akhenaten II* (Egyptian Exploration Society, 1942)

Gosse, Edmund. *Sir Henry Doulton,* edited by Desmond Eyles (1970)

Harington, Sir John. *The Metamorphosis of Ajax* (1596)

Hatton, Joseph. *Twyfords, a Chapter in the History of Pottery* (1905)

Hawkley, B. 'A History of the Ball Valve', *Plumbing Trade Journal*, September 1961, ff

Hellyer, S. S. *The Plumber and Sanitary Houses* (5th edn 1893)

Hellyer, S. S. *Practice and Principles of Plumbing* (2nd edn 1893)

Hetherington, J. R. *Selina's Aunt* (Birmingham, 1965)

Hibbert, C. *London—the Biography of a City* (1969)

Hitchcock, H. R. *Early Victorian Architecture in Britain* (1954)

Kelly. *Post Office Directory for Monmouthshire and the South West* (1871)

Lamb, H. A. J. 'Sanitation—an Historical Survey', *Architects' Journal* (1937)

Lewis, R. A. *Edwin Chadwick and the Public Health Movement, 1832–1854* (1952)

Maddock, A. M. *The Polished Earth: a History of the Pottery Plumbing Fixture Industry in the U.S.* (New Jersey, 1962)

Maddocks, Thomas & Sons Co Ltd. *A History of the Pottery Industry and its Evolution as applied to Sanitation in the U.S.* (1910)

Middleton, G. A. T. *The Drainage of Town and Country Houses* (1903)

Midwinter, Eric. *Old Liverpool* (Newton Abbot, 1970)

Moore, E. C. S. *Sanitary Engineering* (1901)

Newton, Robert. *Victorian Exeter* (Leicester, 1968)

Norton, F. H. *Fine Ceramics: Technology and Application* (New York, 1970)

Paveaux, Andre. *Daily Life in England in the Reign of George III* (1969)

The Pictorial Gallery of Arts (1855)

Pigot. *Commerical Directory for 1829*

Pudney, J. *The Smallest Room* (1954)

Reyburn, Wallace. *Flushed with Pride. The Story of Thomas Crapper* (1969)

Reynolds, Reginald. *Cleanliness and Godliness* (1943)

Searle, A. B. 'Stoneware and Sanitary Ware', in *Industrial and Manufacturing Chemistry*, by G. Martin (1954)

Stephens, W. B. (ed). *The History of Congleton* (Manchester, 1970)

Walters, Stanley. *Mrs Astor's Horse* (New York, 1960)

West, Rebecca. *Grey Lamb and Black Falcon* (1942)

Wright, Lawrence. *Clean and Decent* (1960)

TRADE CATALOGUES CONSULTED

Adamsez
Armitage Ware
Bostel Environmental Systems
British Sanitary Fireclay Association
Dent & Hellyer
Doulton & Co
Duckett J. & Sons
Dudley, Thomas Ltd
Fireclay Manufacturers Association
Flushing Cistern Manufacturers Association
Humpherson & Co
Johnson, Alfred & Son Ltd
Johnson Bros (Hanley) Ltd
Shanks & Co Ltd
Shaws Glazed Brick Co
Twyfords
North Western Ceramic Pipe Co
Valson Enamelled Sanitary Fireclay Co
Plumbing Equipment News
Plumbing & Heating Journal

Acknowledgements

WITHOUT the assistance of other people and other writers a book like this could never be written. I acknowledge with thanks help from the following: Mr E. G. Ellis of Twyfords; Mr W. Pidgeon of Humphersons; Mr J. G. Forsyth of Shanks; Mr R. T. Buck of Dent & Hellyer; Mr W. G. Wright, formerly of Armitages; Mr K. D. Miller, City Librarian, Hanley; Miss E. Hackett and the staff of Monmouth Library; the Curator and staff of the Castle Museum, York; Mr J. Johnson, formerly of Ducketts Ltd; and Mr Ronald Harris of Newport.

I wish to mention also Mr A. D. F. Shanks, of Armitage-Shanks, for permission to photograph the ancient WC bowls at Armitage.

For permission to quote from previously published sources, I gratefully acknowledge: Batsford Books Ltd, for extracts written by G. A. T. Middleton and S. S. Hellyer; G. Bell and Sons Ltd, for extracts written by S. S. Hellyer; and E. & F. N. Spon Ltd, for extracts written by P. J. Davies.

Additional thanks and acknowledgement are due to: Architectural Press for permission to reproduce three drawings from the *Architects' Journal*; the Egyptian Exploration Society for permission to reproduce a photograph of a wc seat at Tel-el-Armana; and Harrods Ltd for permission to publish a photograph of part of their Ladies' Room.

For technical help I am indebted to the *Plumbing & Heating Journal* whose services were much appreciated. I am also grateful to the North Western Ceramic Pipe Co for the gift of two of Ducketts' catalogues of 1910.

Index

It has not been thought fit to include in the index the name of every WC or pattern. Most of these will be found in Chapter Six. Only those mentioned elsewhere in the text occur in the index.